T0220709

Research Methods for Engineers

Learn how to plan for success with this hands-on guide to conducting high-quality engineering research.

Plan and implement your next project for maximum impact
Step-by-step instructions that cover every stage in engineering research, from the identification of an appropriate research topic through to the successful presentation of results.

Improve your research outcomes
Discover essential tools and methods for producing high-quality, rigorous research, including statistical analysis, survey design and optimization techniques.

Research with purpose and direction
Clear explanations, real-world examples and over 50 customizable end-of-chapter exercises, all written with the practical and ethical considerations of engineering in mind.

A unique engineering perspective
Written especially for engineers, and relevant across all engineering disciplines, this is the ideal book for graduate students, undergraduates, and new academics looking to launch their research careers.

David V. Thiel is a Professor and Deputy Head (Research), Griffith School of Engineering, Griffith University, Australia. He has been teaching engineering research methods to students for several years, has managed numerous industry research and development contracts, is the author of over 120 papers published in international journals and is a Fellow of the Institution of Engineers Australia.

Research Methods for Engineers

David V. Thiel
Griffith University, Australia

CAMBRIDGE
UNIVERSITY PRESS

CAMBRIDGE
UNIVERSITY PRESS

University Printing House, Cambridge CB2 8BS, United Kingdom

Cambridge University Press is part of the University of Cambridge.

It furthers the University's mission by disseminating knowledge in the pursuit of education, learning and research at the highest international levels of excellence.

www.cambridge.org
Information on this title: www.cambridge.org/9781107610194

© Cambridge University Press 2014

First published 2014

A catalogue record for this publication is available from the British Library

Library of Congress Cataloguing in Publication data
Thiel, David V., author.
Research methods for engineers / David V. Thiel, Griffith University, Australia.
   pages   cm
Includes bibliographical references and index.
ISBN 978-1-107-03488-4 (hardback) – ISBN 978-1-107-61019-4 (paperback)
1. Engineering – Research – Methodology.   I. Title.
TA160.T45   2014
001.4′2 – dc23   2014007601

ISBN 978-1-107-03488-4 Hardback
ISBN 978-1-107-61019-4 Paperback

Contents

Preface

This book is unashamedly idealistic. It aims first to convey to engineers and engineering undergraduates interested in conducting research a reminder of the fundamental principles of engineering, and then to explain the requirements of conducting excellent, publishable research which will benefit humankind. A number of issues distinguish engineering research from other forms of scientific research. These issues include the dedication of engineering to the betterment of humankind, an acknowledgement of the codes of ethics regulating all engineering activities, the use of engineering standards to ensure quality and acceptable research outcomes and a conviction that sustainability is now a major engineering imperative.

The book is based on a lecture course delivered at Griffith University in the engineering school for coursework master's degree students both on and off campus. The initial concept for the course was developed by Professor Sherif Mohammed at Griffith University. This was revised substantially by the author and then further modified by other faculty staff involved in delivery of the course. The course has also been substantially improved by the students themselves – many of these students have English as a second or third language. To accommodate this, the book contains many starting hints particularly aimed at these students, so that students can rapidly progress without resorting to direct copying from other sources.

The course required the students to submit three written assignments using engineering journal format (a template was provided) and appropriate structure and language. All three assignments were based on a published journal research paper selected by the student individually. The first assignment required the students to write a literature review based on the journal paper and additional relevant papers published more recently. The second assignment was a summary of the research methods used in the selected paper, and was combined with the Assignment 1 literature review, but modified in line with the instructor's feedback. The third assignment built on the previous two assignments and instructor feedback, but had the additional requirements of including a research plan, the research team, data analysis and data presentation. The students were required to create an ideal outcome graphically and to describe the statistical analysis to be used to verify their conclusions resulting from their research outcomes.

The on-campus version of the course involved two classroom hours per week – one lecture and one workshop. Students were required to prepare for the workshop using topics presented to them before the class. Their workshop sheets were submitted at the end of each workshop for feedback. All workshop sheets were directly relevant to the preparation of their assignments. Commonly students were required to comment on each other's work, to present their work and to modify their workshop sheets in line with the class and lecturer feedback. Some of the exercises in this book were used for this purpose.

The off-campus course was run in a similar manner, with students having access to the recorded lectures and the Powerpoint presentations from the lectures. Students were also required to prepare the same workshop sheets, and comment on each other's work using email. The feedback from the instructor and other class members was used to guide the students to improve their methodology and written work for their next submission – either the next workshop or the next assignment.

For both on-campus and off-campus delivery, the student feedback was excellent. As with all courses, the more effort that students put into working with the material, the greater the rewards. The course taught even those students who were not aiming to undertake research work or a research career to look for academic research rigour in the published articles they read, and so to distinguish between solid research outcomes and other reports. It also provided guidelines for report writing – a common task in commercial and industrial workplaces. These skills are beneficial to an engineering career.

All engineering disciplines can engage with the material in the book as the students self-select papers to review which are relevant to their specific discipline. Readers will quickly note that many of the exercises rely heavily on a fundamental understanding of a student's particular engineering discipline. Students who took the course without this fundamental knowledge of a discipline struggled significantly.

I always enjoyed giving these classes. I hope readers enjoy this book and can make a positive contribution to humanity through their engineering expertise and research outcomes.

In conclusion, I would like to thank my fellow academics in the Griffith School of Engineering and the students who have provided me with excellent motivation to teach the course. Most importantly, I express my thanks to my lovely family who have always provided me with wonderful support.

David Thiel

An introductory note for instructors

This course was run at master's degree level at Griffith University for engineering graduates. For convenience, the book refers to the intended readers of the book as novice researchers, regardless of their status. While the first course was delivered to graduate students, some universities have introduced a research methods course in their engineering undergraduate degree programmes. When teaching the course, we assume the following knowledge gained from undergraduate engineering studies:

- A basic understanding of the fundamental concepts and language in a relevant engineering discipline;
- Some experience in laboratory experimentation including the application of mathematical laws to plot and understand sets of results;
- A basic understanding of measurement theory, errors in measurements and statistics;
- The efficient use of a method of analyzing and plotting data (e.g. Matlab or MS Excel).

As a number of the topics in the book are covered in undergraduate engineering degree programmes, the content of this book provides a concise introduction to these concepts. The reader should look to more comprehensive texts for a more careful, detailed analysis or to gain an understanding of the scientific background behind the use of these techniques. Each chapter includes a small list of references. More importantly, readers are given some keywords to conduct searches for further information on any topic.

Many of the techniques outlined in the book are very quick and simple to implement using electronic tools. It is a five minute task for students:

- To plot a set of experimental data points as points and to include the line of best fit (usually a straight line), calculate the equation and the correlation coefficient;
- To find published scientific papers using a keyword search for a literature review.

A failure to do these very simple things in a research report, thesis or research paper suggests to the reader that the author is not a competent researcher. Novice researchers must gain good habits if they are to become efficient, rigorous members of a research team.

Undergraduate engineering students can be taught good experimental and writing habits if the classroom materials are presented in a professional manner. The class laboratory notes should be supplied in the form of a research project outline (see Figure 3.1) complete with an introduction, background theory, measurement techniques and suggested data analysis. There should be adequate references to the textbook and other published papers. Instructors can set an excellent example through this approach, and also demand a similar level of reporting.

1

Introduction to engineering research

1.1 Why engineering research?

The disciplines of engineering are all described as the application of science to realistic systems which benefit humankind [1]. Engineering research is therefore based on the principles of scientific research which, in turn, are based on the scientific method, in which observations (experiments), theories, calculations and models are derived from the existing body of scientific knowledge and verified independently by others who are experts in the field [2–4]. This latter process is called 'peer review'. While this formal review by peers is not foolproof, it constitutes the best method of validation and verification of research results. Engineering research is based on precisely the same scientific method; however, the research is directed toward the practical application of science to products, services and infrastructure.

Most research starts with a hypothesis; that is, a statement which can be either proved or disproved. In most cases it is easier to disprove a hypothesis because only one counter example is required to discredit the idea. To prove a hypothesis, it is necessary to exhaustively examine every possible case and make sure the hypothesis applies. Often this results in the creation of limiting conditions. The conclusion becomes slightly modified in that the hypothesis is valid providing certain conditions are met. A full evaluation of a hypothesis may take many years without a conclusive resolution.

Example 1.1 Hypothesis statements

'All mechanical systems can be described by damped simple harmonic motion equations.' You could test many mechanical systems and find that this is true. However, if you find one example where this is not true, then the hypothesis fails. In this case, it is necessary to apply some limits to the statement.

'The maximum efficiency of a solar cell is 28%.' If you find one example of a solar cell which has a higher efficiency, then the hypothesis fails.

'The laws of physics apply throughout the universe.' Physicists and astronomers continue to assume this is true when describing the formation of stars.

The history and philosophy of science encapsulates the scientific method and the creation of new knowledge [2–4] based on a new theory which has been subsequently verified by observation, experimentation and the logical development from previously accepted theories, but this is not the subject of this book. In some cases theoretical concepts are deduced long before experimental verification. In other cases, theoretical concepts are deduced from experimental observations. The history of science is full of examples of both.

There are many books which discuss scientific research and its methods [5–8]: so how does engineering research differ from research in science? A preliminary answer is to require engineering research activities to fulfill all of the following objectives:

- The research must be applied to human systems;
- The research must yield practical outcomes;
- The research must yield outcomes which benefit humanity;
- The research must be ethically based;
- The research should consider environmental outcomes;
- The research must be based on standard industry based testing.

A more detailed explanation of these issues is provided in the following chapters.

Example 1.2 **Research opportunities**

A new material has been proved scientifically to be a reliable replacement for asphalt and concrete for road building. The material has excellent physical and chemical properties. An engineering research study of this material might also verify that the material is in abundant supply from renewable resources, the material itself can be recycled at the end of its usefulness in road material, and the toxicity of the material does not adversely impact the environment.

A new transistor technology is based on a rare-earth metal which has extremely low abundance, is difficult to obtain and difficult to recover from e-waste. The research in this field constitutes esoteric science rather than engineering research because the outcomes are unlikely to be adopted widely unless improved environmental outcomes can be assured.

Air transport using hydrogen balloons requires very little energy to raise the load and return the load to the ground. It was found to be not practical because of the flammability of the gas, its confinement/storage is difficult and the speed of movement is highly limited.

Clearly scientific research and engineering research are not mutually exclusive. All medical science is directly related to improving the human condition through medical practice. Other human related fields, for example physiology, pharmacy, dentistry, psychology, education, etc also have some outcomes relevant to improving the human condition. Similarly many engineers engage in purely scientific research to test ideas with the long term aim of finding solutions to the practical implementation problems associated with the research outcomes. As there is no clear dividing line between these fields, many conferences and scientific journals report both scientific and engineering based research. This book is confined to engineering based research strategies, but the concepts are also very applicable to purely scientific research. Thus, further reading is readily available from scientific research methods books and papers [7–9].

Engineers, and hence engineering research, are constrained by quite formal codes of ethics. Every discipline of engineering has a code of ethics covering engineering practice in one or more different countries. The codes should influence how the research is conducted and ensure that the outcomes are improvements to humankind through improved environmental outcomes and minimal risk to users of the technology. In particular the concept of economical engineering design must be balanced with aspects of fail-safe and an understanding of materials and product reliability. In many cases these aspects are inherent in the research design, but good engineering research outcomes will require independent verification of not only the research itself, but also the applicability of systems designed from these research outcomes. Codes of ethics and their importance are discussed in Section 1.4.

A research project is not complete until the results have been presented publicly for other experts in the field to comment and review. Thus publication of findings in the open, refereed, international literature and/or presentation at a meeting of research peers is an essential requirement of any research project. Only when the research outcomes have been reviewed by suitably qualified peers can the researchers declare that new knowledge has been created. This means that work conducted in secret (for example in a military research facility, in a high security research laboratory, or in other private venues), does not contribute to the world-wide body of knowledge, and therefore cannot be described as research.

Example 1.3 **Unsubstantiated claims**

Claims that top secret research by the US Government had revealed the existence of unidentified flying objects and the landing of extraterrestrial creatures have never been subjected to international scrutiny and so must not be regarded as contributing to new knowledge.

Some claims of aromatherapists, chiropractors, water divin-
ers, etc have never been substantiated by rigorous scientific
examination and so do not contribute to new scientific or
engineering knowledge.

Similarly, a search of previous publications and patents does
not constitute research. Thus, when a primary school child conducts 'research' on the Great Wall of China by copying the outcomes located using a computer search engine, this does not constitute rigorous scientific or engineering research. This student is
gathering well established and previously reported information.
This is an important distinction: research outcomes which are new
to the researcher but are well known to others does not constitute
original, publishable research. As a logical consequence of this
argument, any original research must clearly identify all relevant
prior work before the authors can claim to have developed new
knowledge. This can be a significant challenge as the volume of
published works continues to grow at an accelerating rate.

1.2 Next step research

All research is built on the background and understanding of science developed over the centuries. When a person plans to engage in a research project to create new knowledge, it is vital that a recent and thorough understanding of the field is gained before designing the research project. A new research project will be built on the work of others, from Newton and Maxwell to Mohr and Edison. In addition, the research strategy and methods applied should be well regarded by the world-wide community of scholars.

In order to emphasise this concept, it is possible to describe two approaches to research: (a) A new fundamental innovation that changes the way we think about the world in scientific terms. This can be described as a paradigm shift. (b) A step forward in our understanding of the engineering world based on one or more of the following ideas:

- The application of techniques commonly used in one field to another field;
- The modification of an existing concept or technique with improved outcomes;
- The modification of current technologies for improved efficiency, miniaturization, sustainability or environmental outcomes.

Example 1.4 **Translational research opportunities**

Mechanical engineers used the finite element method for many years before the electromagnetic equations were solved numerically using the method. This resulted in a new field of computational electromagnetics in the 1980s.

Image analysis techniques used for face recognition and satellite based vegetation categorization can be applied to two and three-dimensional data sets in any field of engineering.

Inertial sensors used in the automotive industry as airbag triggering devices are now used in sports engineering for movement analysis and in mechanical engineering for vibration analysis.

1.3 Research questions

A common method of focusing on a research project is to phrase a research question. The design of a single, succinct question is a challenge for all researchers and the research team may consider several iterations before it is accepted. The research question will directly lead to one or more methods of investigation, and these can be divided into a number of research aims.

The research question can be phrased using one of the following questioning words:

Why?

Example 1.5 Research question 'why?'

Why did the wind turbine fail in 100 kph winds?

This question suggests a number of avenues of investigation. For example the researchers might:

- review the literature for previous failure reports,
- assess the wind conditions at the time of failure,
- undertake numerical modelling experiments,
- review fatigue and possible points of weakness,
- conduct inspections of other wind turbines located in the area.

What?

Example 1.6 **Research question 'what?'**

What is the effect on the strength of concrete when recycled concrete is used in the mix?

This question suggests a number of avenues of investigation. For example the researchers might:

- review the literature seeking results from previous trials,
- conduct compression and shear experiments using different mixtures of concrete,
- conduct strength calculations based on aggregate strength theory.

How?

Example 1.7 **Research question 'how?'**

How can the braking system of a railway carriage be self-activating when its velocity exceeds a threshold value?

This question suggests a number of avenues of investigation. For example the researchers might:

- review the literature and patents for automatic braking systems,
- calculate the braking power required,
- conduct model-based experiments on braking systems.

When?

Example 1.8 **Research question 'when?'**

When will the roof bolts in an underground tunnel fail through environmental degradation?

This question suggests a number of avenues of investigation. For example the researchers might:

- review the literature for previous studies in different rock types and environmental conditions,
- review the types of roof bolts in common use,
- conduct a survey of rock bolts in different tunnel environments to assess degradation,
- conduct experiments to measure the degradation of the roof bolts under accelerated environmental conditions.

These examples might suggest some of the work that has been previously reported. A review of the literature will mean that the research team does not have to 're-invent the wheel', and can build their research on the published reports of others. It will also suggest that even if the same problem has not been solved previously, the methods used to solve similar problems might be appropriate to solve their specific problem. A good literature review can impact positively on the research methods that the research team might use. This is of significant value as the use of previously reported and peer reviewed methods adds confidence about the reliability of the research method and the subsequent independent review of the journal and conference papers arising from the research.

Novice researchers should also note that a number of different methods of approach are suggested for each research question. It is mandatory that more than one method of investigation is used in all research projects in the hope that the results from a number of different approaches can be used to substantiate the conclusions from the project. This adds confidence in the research outcomes.

1.4 Engineering ethics

As engineers are involved in all major infrastructure projects (dams, bridges, roads, railway lines, electricity distribution, telecommunications, vehicles, etc), it is not surprising that when one of these facilities fails, caused either by natural events (earthquakes, adverse weather, landslides, material failure, etc) or by human intervention (terrorist attack, stadium overloading, land contamination, water contamination, air pollution, lack of maintenance of facilities, land subsidence due to mining, etc), inevitably some of the blame and responsibility is levelled at the engineers who undertook the design, construction, maintenance and control of the facilities. Over the past 200 years, there have been engineering failures that have resulted in loss of human life, medical problems, the extinction of species, damage to the environment, and damage to the economic wellbeing of towns and entire countries. As a consequence of these disasters, the professionals engaged in engineering activities have created and joined professional societies. These societies are designed to minimize the likelihood of repeat occurrences through member registration and mandated self-improvement. This is achieved through two methods; firstly by the free exchange of information between practitioners, and secondly, through an adherence to codes of ethics which are designed to eliminate poor practice and, in extreme cases, to prevent negligent and incompetent people from placing the community at risk by working on such projects. These professional engineering societies therefore engage in processes of accreditation of individuals and university degree programmes, and the publication of research, engineering standards (best practice) and failure analyses.

Example 1.9 **Engineering disasters**

The mining of phosphate on small islands has resulted in untenable farmland, local climate change and water shortage for the inhabitants.

The draining of a lake in Centre Asia resulted in the destruction of the environment and subsequently village life.

The incorrect fitting of the fuel tank seals in the Challenger space shuttle resulted in the vehicle exploding during launch.

The meltdown of the nuclear reactor at Chernobyl killed many people, caused significant radiation damage to those who survived, and caused the local area to be contaminated for the next several hundred years.

Engineering societies require that practising engineers take a holistic view of projects which deliver a complete solution providing maximum benefit to all stakeholders (e.g. the community, the users of the product or service, their company or organization commissioned to undertake the work, and the environment both local and international).

In most countries professional engineers are registered. It may be mandated in law that only a registered engineer is allowed to build a bridge or sign off on a control circuit used in a theme park ride. Without the signature of a registered engineer, the work is not permitted to start, or not permitted to operate.

The registration process requires the person to be suitably trained (a university degree in engineering is most common) and must pledge to work within the guidelines of a Code of Ethics. Should engineers not conduct themselves appropriately, they can be removed from the register of professional engineers and barred from continuing to practise. Most engineering codes of ethics also prohibit engineers from practising outside their engineering discipline.

While most engineering codes of ethics differ in the fine detail, there are some tenets (or statements) which are common to almost all of the engineering codes [8].

Importantly, professional engineers are required to act in the best interests of humanity and the community. This requirement stands above all else and overrides engineers' responsibilities to:

- Their employers;
- Their family, friends and relatives;
- Their town, region and country.

From this basic tenet, other tenets are derived which relate to:

- The environment (including sustainability);
- Public safety (including fail-safe design);
- The economic viability of their employer (cost-effective, reliable outcomes).

The adherence to a code of ethics maintains the reputation of the engineering profession. For example the Institution of Engineers Australia code of ethics begins with the words:

> 'As engineering practitioners, we use our knowledge and skills for the benefit of the community to create engineering solutions for a sustainable future. In doing so, we strive to serve the community ahead of other personal or sectional interests.' [9]

The American Society of Civil Engineers (ASCE) use the following fundamental principles [10]:

> 'Engineers uphold and advance the integrity, honor and dignity of human welfare and the environment by:

1 Using their best knowledge and skill for the enhancement of human welfare and the environment;
2 Being honest and impartial and serving with fidelity the public, their employers and clients;

3 Striving to increase the competence and prestige of the engineering profession; and

4 Supporting the professional and technical societies of their discipline.'

And the first ASCE fundamental cannon is:

'Engineers shall hold paramount the safety, health and welfare of the public and shall strive to comply with the principles of sustainable development in the performance of their professional duties.'

Many professional engineering organizations are now international, and this allows engineers trained in one country to operate without impediment in another country if both subscribe to a common code of ethics. The largest international engineering agreement is the Washington Accord [11], where signatory nations must allow regular inspection of the academic qualifications of their member nations and adherence to a common code of ethics.

The codes of ethics commonly restrict the unauthorised copying of designs and other intellectual property. For this reason, engineering researchers must acknowledge the work of others in the development of their findings.

When undertaking engineering research, the research team must understand the implications and restrictions which apply to their research projects based on their code of ethics. There is also an obligation that researchers must undertake their work to the best of their ability and in line with their training, within the standards defined by their discipline, to use appropriate terms to describe their work, and to provide unbiased reports on the success of their work [8, 12]. It is most important that the conclusions of a research project not only summarise the outcomes in a positive light, but probe the uncertainties and problems which might occur if the research results are applied to products and services. These matters will be discussed in some more detail in later chapters of this book.

When the research involves the use of humans or animals, research funding bodies and the publishers of research outputs (journals and conference technical committees) require that the project plan be assessed for impacts on the human and animal subjects before the project begins. Such research projects include surveys (the participants are asked to respond to a set of questions), physical activities (the participants are asked to perform manual tasks) and mental activities (participants are asked to solve puzzles). Asking individuals to participate in a research project either voluntarily or for a reward or other benefit, can have significant negative consequences on the volunteers and the research outcomes.

Example 1.10 **A biased survey**

A company decides to ask its employees for feedback on their latest product, which was designed in-house. Some employees feel obliged to support the product concept because they do not wish to offend the design team or the general manager. The net result is a biased outcome of the survey and potential damage to the careers of those employees who did not respond favourably.

In order to avoid the problems of biased research outcomes and perceived or real threats to the participants, an ethical approval procedure must be available to ensure that negative outcomes to the research and the research participants are avoided.

This requires the following information to be presented to the target group in an understandable way:

- The task is voluntary;
- The participants are provided with information about the tasks and risks before commencing;
- The participants are provided with contact details if they have additional questions or feel aggrieved;

- The participants can withdraw at any time during the survey without recrimination;
- The survey responses are anonymous and the data are stored without the identification of individual participants;
- The accumulated results are provided to the participants when the final analysis has been completed.

These requirements can be achieved in surveys using a number of different methods of approach:

- A common method is to have the work (for example surveys can be administered by an independent organization) conducted by an independent third party. This party will solicit the volunteers, collect the responses and ensure that the company seeking the data will not be provided with identification tags for individuals.
- There are now options for online surveys which are effectively anonymous.
- The individual profiles used in the survey are not sufficiently specific so that each individual cannot be easily identified.

The design and conduct of surveys are discussed in more detail in Chapter 6.

Example 1.11 Individual identification in surveys

A participant profile which sought precise values for age, height and weight could easily result in the identification of an individual. A participant profile which sought responses to broad ranges of age, height and weight would reduce the chances of individual identification.

1.5 What constitutes conclusive proof?

Research is designed to create new knowledge. This new knowledge needs to be substantiated appropriately, initially by the research team, and subsequently through the peer review process by the world-wide community of scholars who are experts in this field. Thus novice researchers must convince themselves that their results and conclusions are valid and supported by strong evidence. This is commonly done using more than one of the following techniques:

- Experimental measurement (particularly using standard tests);
- Theoretical development;
- Logic and mathematics;
- Numerical simulation;
- Statistical analysis;
- Comparison with previously published research outcomes;
- The use of multi-parameter optimization methods to obtain the best outcome of the design.

Ideally a research team would engage in many, if not all, of these techniques to verify its results. This gives the independent reviewers of the work the best possible proof that the new knowledge has been validated and the conclusions are correct.

Often, however, there may be problems in using this approach, for instance:

- There are no standard experimental methods applicable to a particular research investigation;
- Repeated experimental measurements are not possible so that statistical support is difficult;
- The theoretical frameworks available are too simplistic to describe the outcomes;
- There is no prior work which closely resembles this investigation; and
- Multi-parameter optimization is not possible as computational models are ineffective.

These difficulties can all result in a probability of failure or project risk assessment. This is usually expressed as a probability.

The publication of research results mandates that some supporting evidence is provided. A failure to do this will result in a failure to have the research outcomes accepted. It is therefore very important that the research team design a project in a manner which ensures that sufficient supporting evidence is available at its completion, either from the work itself or from other published works.

Example 1.12 Questions for probability

A glass window in a high rise building falls to the street below. A structural engineer is asked to calculate the probably that this event will occur again. How can the engineer verify her results given that only one event was recorded?

The flight control system in an aircraft is rated as 99% reliable. Would you fly in this aircraft?

A civil engineer quotes a dam wall as being able to withstand a 1 in 100 year weather event. Would you be happy to live beneath the dam wall?

In all cases, conclusive proof is required for the safety of the population.

In mathematics, the concept of upper and lower limits can be applied to gain information about the solution to seemingly intractable problems. For example, if a function is not integrable (i.e. cannot be integrated using analytic means), one can choose two integrable functions – one of which is always greater than the function, and the other is always less than the function. By completing the two simpler integrals one can deduce the range within which the unknown function lies. In the same manner, simple models which provide upper and lower bounds to a more complex problem can be solved theoretically and/or computationally to deduce the likely range of outcomes from the more complex model. This method can provide additional support to the research results, even when a complete model has not been solved.

1.6 Why take on a research project?

There are many reasons for an engineer to become involved in a research project. There are benefits to society as part of an engineer's charter as well as personal rewards.

There is significant excitement in new discoveries. The possibility of developing something completely new, something of benefit to humanity and something to add to the world-wide body of knowledge is a strong motivation for undertaking research. While practising engineers may leave their mark on society through buildings, dams, rail-lines, aircraft, electronics products, medical devices, etc, usually the team of engineers or their company might only be known to the general public through a temporary sign on the work-site or a slim column in a newspaper. Sanitation and pipeline engineers, software engineers and many others might never be recognised publicly. It is often the case that the names of engineers become known when there are catastrophic failures. The publication of research outputs in the archival literature means that names of the members of the research team will never be lost in time. The research team will be recorded permanently.

A successful research project can enhance your career. The peer review process and open publication means that the team is capable of work at the highest possible standard. This recognition is world-wide and can be used to advantage in developing tenders for international projects.

The process of research training is part of every engineering undergraduate degree programme. Every laboratory experiment and every calculation is part of the development of an engineer, and the skills learnt are part of the expertise that can be used in research projects. For this reason many undergraduate engineering degree programmes either include or plan to include a research project or a research methods component. Undergraduate research training can provide a thorough grounding in many of the techniques outlined in this book.

In a complementary way, engineering faculty staff should mandate that laboratory reports, assignments and other deliverables employ the typical research requirements outlined in this book – the research writing format, statistical analysis, appropriate checks on the validity of results and appropriate referencing. In addition, the development of lecture material and other resources should follow the norms of research methods, referencing, etc. This solid reinforcement of research techniques can greatly improve the quality of graduates and improve the student experience in undergraduate engineering programmes.

1.7 Chapter summary

The objective of all engineering research is the creation of new knowledge which can be of benefit to humanity. New knowledge is based on well-known and well-accepted principles of scientific thought and measurement, and should be initiated through the development of a research question.

In order that a research contribution can be claimed, novice researchers must have a good understanding of the basic principles of their discipline as commonly found in the well-known textbooks in the field.

All engineering research is based on the scientific method of validation using the peer review process. Peer review of the research outcomes is an essential process in research.

All engineering researchers must be familiar with their code of ethics and must behave ethically. In particular the research team must seek out and acknowledge the work of others.

Exercises

Many of the exercises in this book require the reader to choose a refereed, published paper in their discipline, to analyse the paper according to the instructions given in the exercises, and to extend the work reported as a new line of research.

1.1 Use an academic web search to locate a journal paper which describes a design outcome in your field of interest (i.e. your engineering discipline). You must enter several keywords which relate to your topic. Read the paper and, using your own words, demonstrate your understanding of the paper by:

- Writing out the major conclusions of the paper;
- Outlining the verification method(s) used to support these conclusions
- Describing the authors' reflective comments on the quality of the design (positive and negative).

List your comments on the following:

- The positive and negative environmental impacts of the new design;
- The fail-safe quality of the new design;
- The cost of manufacture or implementation of the new design compared to previous designs.

1.2 Find the code of ethics which covers your engineering discipline (the code might be specific to one particular country or geographical region). Rank the following aspects in terms of importance (usually indicated by the position in the code of ethics). Assume that the first tenet is the most important:

- Responsibility to the environment;
- Responsibility to sustainable outcomes;
- Responsibility to an employer;
- Responsibility to the general public;
- Responsibility to the nation.

Following the strategy in Exercise 1.1, choose a published paper from your engineering discipline and comment on any of these priorities identified in this paper. Compare your results with the code of ethics list. What is your conclusion about the ethical approach taken in this paper?

1.3 After reading a published research paper, write down the research question you think the authors have addressed in undertaking this research. Do you think the paper adequately supports the conclusions reached in addressing this question?

1.4 In your undergraduate education or master's degree, choose three experiments (practical laboratory work) and identify the following:

- The aims of and methods used in the experiment;
- The conclusions from the experimental work;
- The methods of verification used to support your conclusions;
- The theory applied to support the experimental outcomes;
- The statistical techniques used to verify your outcomes.

Do you consider that the post experimental analysis supported the experiment aims and conclusions? Outline how this research method might be improved to verify the outcomes through conclusive proof.

References

Keywords: engineering research, engineering ethics, code of ethics, research question, engineering sustainability, engineering disasters

[1] Davis, M., 'Defining engineering: how to do it and why it matters', in M. Davis (ed.), *Engineering Ethics*, Aldershot, England: Ashgate, 2005.

[2] Popper, K.P., *The Logic of Scientific Discovery*. New York: Basic Books, 1959.

[3] Gattei, S., *Karl Popper's Philosophy of Science: Rationality without Foundations*, New York: Routledge 2009.

[4] Kuhn, T.S., *The Structure of Scientific Revolutions*, (2nd edition), Chicago: University of Chicago Press, 1970.

[5] Marder, M.P., *Research Methods for Science*, Cambridge, UK: Cambridge University Press, 2011.

[6] O'Donoghue, P., *Research Methods for Sports Performance Analysis*, Oxford, UK: Routledge, 2010.

[7] Walliman, N., *Research Methods, the Basics*, Oxford, UK: Routledge, 2011.

[8] Fleddermann, C.B., *Engineering Ethics*, (4th edition), Upper Saddle River, NJ: Prentice Hall, 2011.

[9] Engineers Australia, 'Our code of ethics', adopted 28 July 2010, www.engineersaustralia.org.au/ethics, accessed 4 December 2012.

[10] American Society of Civil Engineers, 'Code of Ethics, Fundamental Principles', http://www.asce.org/Leadership-and-Management/ Ethics/Code-of-Ethics/, accessed 4 December 2012.

[11] Hanrahan, H., 'The Washington Accord: past present and future', IEET accreditation training, International Engineering Alliance, 2011. http://www.washingtonaccord.org/washington-accord/ Washington-Accord-Overview.pdf

[12] Whitbeck, C., *Ethics in Engineering Practice and Research*, Cambridge, UK: Cambridge University Press, 1998.

2

Literature search and review

2.1 Archival literature

The world's total knowledge in the fields of science and engineering is stored in written form as published books and papers. Much of this is now stored digitally and available on-line. For a research team to successfully undertake new research they must contribute 'new knowledge' to this total store of knowledge through publication in the same way (i.e. through writing books and papers). In order to assess if a contribution is new knowledge, the research team must take the following steps:

- Review this vast store of knowledge;
- Conduct research to develop additional knowledge by building upon this previous knowledge; and
- Make their new knowledge available to the world-wide research community through publication following a rigorous peer review.

The world-wide published scientific literature is commonly referred to as 'archival literature', because it is permanently stored and is deemed to be of value to future generations of research scientists and engineers. Once information is printed on paper, the content cannot be changed. This form of publication is different from some web based publications, where the content can be changed relatively easily.

Publication in the archival scientific literature should be the common goal of all scientific and engineering research. Once published in this readily available form, engineering research and innovation can be used by all people, both nationally and internationally, for the betterment of human kind and improving the human condition. This is the goal of all engineering endeavours including research activities as outlined in Chapter 1.

There is some scientific literature which is deemed to be worthy of preservation and some which is not. The distinction between these two categories is drawn on the basis of the scientific method and peer review, although the distinction can be a little blurred. A research team must understand and maintain current knowledge about innovations in the field of their research. It is important, however, that novice researchers distinguish between literature that is acceptable to the scientific community and that which is not acceptable. While there are exceptions to the broad publication categories, this section gives some general characteristics and guidelines for research teams when conducting a search in the scientific literature.

The archival literature remains unchanged and available forever. This places great responsibility on researchers to ensure that their contributions are new, valuable and rigorously developed. Failure to do this can have significant negative impacts on the careers of the researchers concerned. Academic rigour in research is also an ethical requirement. Academics at tertiary institutions and researchers at research laboratories experience workplace pressure to publish large numbers of high quality papers in refereed journals. From time to time, a minority of researchers are exposed when their research papers are proved to be not new or the results have been fabricated. This constitutes a violation of the engineering code of ethics.

2.2 Why should engineers be ethical?

There are many answers to this question; some are carrots (= incentives) and others are sticks (= threats). In this section, the first paragraphs directly appeal to professional conduct. The later paragraphs outline the problems that can arise if a researcher does not follow the laws of the community and the rules of the profession.

Let's start with some carrots. As a member of a profession, an engineer is a highly respected member of society. There are many statistics derived from surveys of the general population which show that the engineering profession is one of the most trusted occupations in society. As part of the profession, an engineer is encouraged, through solidarity, to behave ethically like all other engineers, both predecessors and colleagues. In professional employment, engineers must make decisions based on facts and models irrespective of political and social influence. The same is true for engineers engaged in research. Some 'whistle-blower' engineers, i.e. engineers that report oncoming problems irrespective of their company or government bosses, have suffered significantly in their professional life by using the media or going to independent authorities to report malpractice. However, most engineers prosper when they provide independent expert advice to clients and the general public.

A failure to behave ethically can result in some very undesirable effects on an engineering career. In its most simplistic form, copying another person's designs or theories is theft. That is, in most countries, an offender can be taken to a court of law, and if found guilty, can be fined and imprisoned.

A commonly accepted legal definition of a breach of copyright is when more than 10% of a work is copied. In academia this is referred to as plagiarism and the penalties can range from a minor penalty (e.g. failing a particular assignment or examination) to a more major penalty (cancellation of enrolment and dismissal from the university).

Example 2.1 **Legality and unethical practice**

Take a book, cross out the author's name and replace it with your own. Do you think this is legal?

Take the architectural plans for a house, cross out the designer's name and insert your own and sell it to a client. Do you think this is legal?

Such behaviour is intolerable in the engineering profession and will result in legal action.

A wider problem occurs if you copy material without first checking the validity of the calculations or the design. Assuming the design is implemented and a catastrophic engineering failure occurs, the consulting engineer who copied the work, or at least part of the engineering team, will be blamed for the consequences of this failure. Again prosecution can result, to the detriment of careers and future work contracts in engineering. The various professional engineering associations will find such behaviour unacceptable and the people concerned might lose their professional recognition as engineers and be disallowed from practising engineering.

The fact remains that all engineers and publication officers (including the reviewers of scientific research papers) are humans and can make mistakes. This might include mathematical errors, incorrect observations, experimental errors, misinterpretation of data, etc. Should a person accept and copy information without acknowledgement and without exercising professional engineering judgement, these errors can be propagated through the literature to the detriment of the advancement of engineering science. All professional engineers must critically review information before accepting it and incorporating it into a design. If the information is proved to be valid, then the originators of the idea(s) must be mentioned (i.e. cited) as a primary source of the information, and it must be acknowledged that this original information has been copied in part or in full or adapted to suit the current problem. By doing this, the design team of engineers have behaved ethically, in addition to providing some protection should the design have fundamental or structural flaws. In the event of a failure, the team can responsibly say that the initial error was generated elsewhere. This will not negate the responsibility for not finding the flaws, as the team has still failed to adequately question and test the principles behind the design. This is an indication of engineering incompetence and an admission of liability.

In addressing these issues, every engineer must have a strong and fundamental knowledge of their chosen discipline. If there are gaps in this knowledge, then the responsible engineer must call upon expert advice to cover any weakness or uncertainties in theoretical knowledge or understanding.

For almost every article published in a scientific or engineering research journal, the authors or their institution/company are required to sign a copyright declaration passing ownership of the article to the journal publishers or the professional society which publishes the journal. This means that the researcher, anyone in the research team, or any other person, is not allowed to reproduce the words, tables or diagrams/figures in the article in any new publication. This is both an ethical and legal responsibility of the authors, and failure to adhere to this policy can result in the researchers being banned from publishing in that journal and even legal action against the authors found in breach of copyright. This point was discussed in Chapter 1.

Another ethical issue is the proper recognition of the contribution of colleagues in a research team. The researchers who have contributed more than 10% of the research effort and can competently describe and present the research outputs to others, should be included in the list of authors. Thus a research project which required the expert services of a statistician, analytical chemist, materials scientist, etc must ask these questions before including or excluding their names on papers and books submitted for publication. Many journals in the life sciences ask all of the named authors to sign a form verifying their participation in the research project and to indicate the percentage contribution from every author. One contentious issue is whether the head of the research group, the leader of the laboratory, or the person who funded or attracted the funds for the research, should be automatically included in the list of authors. This must be considered before the work is undertaken, rather than when the publication is in the final draft of preparation.

So what happens if the author(s) or publisher makes a significant mistake in the paper, and the paper is published with the error. The members of the research team are obliged to publish a correction to the original manuscript. This should normally be done quickly and, where possible, the correction should be published in a subsequent issue of the same journal as the original article. Should the research team find errors in previously published papers, then writing a short note to the journal editor will allow the authors of the original article to provide feedback and an erratum if appropriate. The publication process is primarily aimed at contributing to the total body of knowledge, and the elimination of errors in recently published articles is of significant value to the profession and the community at large.

2.3 **Types of publications**

Whatever you say
May fade away
Whatever you write
Might come back and bite

While the spoken word, if unrecorded, may not be recalled clearly or exactly, the written word can last forever – whatever is written can and will be used to evaluate the competence of the author. In the technical areas, the works of Newton [1] and Maxwell [2] can be found and read by today's scholars. In an electronic age, where written work is recorded digitally and accessed publicly, papers can circulate the world within a second of publication and the contents will remain available and accessible forever. The written word will define a researcher's technical competence. It is therefore very important that researchers write well, in addition to understanding the importance of the written works from others. While some journals only publish electronically, the paper or electronic version is fixed at the time of publication and cannot be altered. The electronic file format used is the *.pdf (portable document format) which is difficult to edit. Errors in published papers are corrected using a correction note in the same journal that published the original work.

All science and engineering is based on the current understanding of the laws of nature (i.e. the physical, chemical, biological and mathematical laws). Researchers must read, understand and engage with the previously published work of other researchers

Example 2.2 Dismissal resulting from unethical practice

A number of university presidents and high ranking politicians have been forced to resign or have been dismissed from their positions because it was proved that some of their early student work was directly copied from others without proper acknowledgement (see the work of Gutenplag Wiki (Theodur zu Guttenberg, Silvana Koch-Mehrin, Veronica Sas), and others (Moon Dae-sung, Pál Schmitt, Madonna Constantine, Ward Churchill to name a few).

Preserving your academic reputation is very important for your career. If you use the work of others without referencing the source, and the works you plagiarised were wrong, the profession will judge you as either a criminal (stealing ideas from others without acknowledgement) or incompetent (incapable of arriving at correct results from investigations).

in the same discipline of engineering. Note that some of the early published works, say in the 1930s, contain some of the concepts that have since been shown to be incorrect in the light of more recent developments.

Example 2.3 A changing paradigm

Greek philosophers thought light consisted of a stream of tiny particles (corpuscular theory). Newton supported the idea but Huygens advocated a wave theory for light. Young and Fresnel explained interference effects using this wave theory. Light was regarded as a transverse wave through 'ether' based on the view that there was no such thing as completely empty space (we call it a vacuum) through which light could pass. The ether was required to carry the wave. Quantum theory has now replaced all of these theories. [3].

TABLE 2.1 Broad categorisation of publications and their characteristics. A tick indicates that the descriptor is valid. A cross indicates that the descriptor is not valid. The symbol √/× indicates that the descriptor is sometimes valid.

	Journal articles	Conference papers	Books	Standards	Patents	Theses	Trade magazines	Newspaper articles	Infomercials	Advertisements	Wikipedia	Web sites
Archival scientific literature	√	√	√	√	√	√	×	×	×	×	×	×
Evidence of peer review	√	√	×	√	√	×	×	×	×	×	×	×
Author's names	√	√	√	√	√	√	√/×	√/×	×	×	√	√/×
Author's affiliation	√	√	√	√/×	√	√	×	√/×	×	×	×	×
Author's contact details	√	√	√	√/×	√	×	×	×	×	×	×	×
Title word count >10	√	√	×	√/×	√	√	×	×	×	×	×	×
Abstract	√	√	√/×	√/×	√	√	×	×	×	×	√	×
Keywords	√	√	×	√/×	√	√/×	×	×	×	×	√	×
Reflective assessment of results	√	√/×	×	√	×	√/×	×	×	×	×	×	×
References	√	√/×	×	√/×	√	√	×	×	×	×	√	×
Publication date	√	√	√	√	√	√	√	√	×	×	√	√/×

The challenge for researchers is to distinguish what is credible scientific evidence from what is opinion, speculation, and sometimes factual error. This can be done, in part, by checking where the information was published. The strength in the scientific method is peer review. If papers have not been peer reviewed in a rigorous manner, then the findings are suspect.

Novice researchers should distinguish between the various types of publications listed in Table 2.1 and use only those published works that are clearly steeped in the scientific method. The most rigorous publication type is the journal paper. Figure 2.1 exhibits many of the most important characteristics that can be used in defining the validity of the information contained in a publication. While the system of peer review is not foolproof and errors can be found and even propagated in later publications, this is the best and most widely accepted method available and is used to maintain the integrity of science and engineering principles. Researchers must understand this so that they know how and where to publish their work.

Researchers, however, may find many new ideas in most of the publication types listed in Table 2.1. The authors and publication must be cited in any reports if the material has been used as part of the research project. Each publication category given in Table 2.1 is discussed in more detail in the following sub-sections. These sub-sections describe some of the commonly accepted practices but individual publication mechanisms have variations on these processes. The information given in these sections serves as a guide only, and prospective authors must review all information provided by the targeted publication.

2.3.1 Journal articles

Refereed journal articles are the most important and most valued contributions to the archival literature. While the time between the date of submission and the date of publication can be quite long – two years is not uncommon – most journal editors express the wish that the time between submission and publication be as short as possible. A period of six weeks for the return of the first review to the authors is a common target. The delay between submission and publication is caused mainly by the reviewing process. Reviewers perform their reviews on a voluntary basis. For this reason the selection of reviewers can take time as each potential reviewer must be approached (the title and abstract are sent) and then agree to undertake the review process within a set time frame (commonly four weeks). Once the final version of the paper has been approved, then an editorial committee will prepare the paper for publication. This causes additional delays.

Journal articles are subjected to anonymous peer review. This means that two or more experts in the field are required to review the work, suggest corrections and approve the work as scientifically rigorous. While this review process is not always perfect – it depends on the dedication, ethics and competence of the reviewers – it is the best available source of expertise world-wide. Paper reviewers are selected by the journal's technical editorial staff. These reviewers are expected to have the required experience and competence in the field of research.

Once the review has been completed the associate editor will make a decision on the suitability of the manuscript for publication. Commonly the reviewers must decide on one of the following recommendations:

(a) Accept without change;
(b) Resubmit with minor corrections (e.g. correct typographical errors, improve the quality of the graphs, add additional information, add additional references, etc);
(c) Resubmit with major changes (additional research work required);
(d) Reject (insufficient new material, scientific errors, flawed methodology, etc);
(e) Reject (material is not relevant to this journal).

All comments are sent to the authors together with the decision. In the case of categories (b) and (c), the authors must address every point raised by the reviewers, revise the manuscript (showing the changes) and resubmit to the journal for further consideration. The review process will then be repeated, usually with the same reviewers.

In addition to commenting on the research methods, results and conclusions, the reviewers are asked to review the reference list and to provide the authors with a list of typographical errors. Reviewers will also comment on the quality of the graphs and other figures and the relevance of the paper to the journal for which it has been submitted. In the light of these comments the authors revise and resubmit the paper. Further comments will then be made by the reviewers. The paper will only be accepted for publication if all the reviewers are happy with the final draft of the paper. The paper is then passed to the editorial staff of the journal for a further review and additional questions might be asked of the authors. These are questions of clarity and language use rather than technical issues.

Research journals in science and engineering, like most magazines, are published regularly in a yearly cycle. The volume number is usually the same for articles published during the same calendar year. Each issue published in the year contains a number of articles. The issue number is simply a count from the first issue release in the year (issue number 1). The page numbers run sequentially through the year. Thus the first article in issue number 1 starts on page 1. The first article in issue number 2 will continue from the last page number of first issue.

Journal publications commonly report the latest scientific and engineering developments, although some review articles are published in order to consolidate the current knowledge in the field. Review articles might not contain new knowledge created by the authors, but will contain a very large reference list which covers the latest developments. Some journals specialize in this type of article (e.g. *Proceedings of the IEEE*). In both cases the papers are reviewed independently before publication is approved. All scholarly journals use an ISSN number (International Standard Serial number) and this can be found in the front pages of each issue. While most journal names are unique (world-wide), it is important for authors to use the correct journal title as many journals have similar names.

Figure 2.1 shows a typical first page of a journal article. The major features revealed on the front page include:

- The title of the journal;
- The title of the paper (usually containing more than ten words);
- A list of the authors, their affiliations and contact details;
- An abstract which outlines the work and major conclusions;
- A list of keywords relating to the paper;
- The submission date, revision date and the publication details (e.g. the journal volume number, issue number, page numbers and date of publication).

The title is sufficiently long to ensure that the contents of the article are uniquely identified. The author's names, affiliations, contact details are provided to allow other researchers to make contact with the research team to ask further questions about the work. The use of a prescriptive title (commonly more than ten words) ensures that the reader can establish the relevance of this work to their own research endeavours. Commonly an abstract is available publicly on the internet at no cost to the reader. There may be a cost to access the full paper unless the reader subscribes to this particular journal. The inclusion of keywords in the paper allows targeted, high speed computer searching. The references cited in the paper demonstrate that the work was based on the work of others in developing the field and allows the reader to undertake further background searches into the field. The publication date establishes the priority of the research work reported and is used by the research community to follow the path of the research development.

Often two or more dates are published with the article. The first date is the earliest date and establishes when the researchers completed their work and the submitted paper was received by the journal. The review of the article is then undertaken and the paper

IEEE TRANSACTIONS ON ANTENNAS AND PROPAGATION, VOL. 46, NO. 6, JUNE 1998 841

Base-Station Tracking in Mobile Communications Using a Switched Parasitic Antenna Array

Stephanie L. Preston, *Student Member, IEEE*, David V. Thiel, *Senior Member, IEEE*, Trevor A. Smith, *Student Member, IEEE*, Steven G. O'Keefe, *Member, IEEE*, and Jun Wei Lu, *Member, IEEE*

Abstract—Base-station tracking in mobile communications benefits from a directional antenna and so requires direction finding technology. A novel technique for electronically directing the radiation pattern of an antenna array employs a directional array with only one active element and three parasitic elements operating near resonance. Three different methods of direction finding are assessed; a coarse angular location method, a precise angular location method assuming one incident beam, and a precise angular location method with multiple incident beams. An array with *n* elements, if used in conjunction with a relatively simple controller, can be used to resolve *n* − 1 signals. This technology can be implemented using both wire and patch antenna-array elements and either linear or circular polarization can be used, lending the technology to applications in both terrestrial and satellite communications systems.

Index Terms—Antenna arrays, mobile communication.

I. INTRODUCTION

IN digital communications systems, it is possible to have a periodic break in the transmission of information without degrading the signal transmitted. This break in transmission can be used to optimize the communications channel. One factor that can greatly improve the channel is the use of a directional antenna system. In this paper, we propose that in between data segments, the mobile transceiver maintains the optimal channel by re-orienting the antenna system. This is only possible if the time required to perform the operation is sufficiently small and the directivity of the antenna is adequate. We suggest that an electronically steerable switched parasitic antenna array supported by a small digital controller can achieve these objectives, even in severe multipath environments. The solutions we suggest are evaluated in the context of time-domain multiple access (TDMA) in group special mobile (GSM) mobile telephone communications for use in base-station tracking but are far more generally applicable.

When using directional antennas, intelligent high-speed direction-finding techniques are required and, in the case of a mobile transceiver, the system must have low-power requirements. With conventional directional antenna systems

Manuscript received March 5, 1997; revised December 29, 1997. S. L. Preston was supported by a scholarship from the CSIRO Division of Telecommunications and Industrial Physics, Sydney Australia. This work was supported in part by research grants from the Australian Telecommunications and Electronics Research Board and the Australian Research Council.

The authors are with the Radio Science Laboratory, School of Microelectronic Engineering, Griffith University, Nathan Qld, 4111 Australia.

Publisher Item Identifier S 0018-926X(98)04873-X.

for initial signal acquisition and subsequent direction updates, a full 360° scan may be required. We propose a technique that will enable the updates to be minimized and allow the direction to be determined without a full 360° scan periodically. Current direction-finding techniques include conical scan, sequential lobing monopulse, and track-while scan [1], [2]. These techniques require the use of either mechanically rotating antennas, crossed loops or phased arrays.

In direction finding, multiple incoming signals can result in an incorrect angular position being determined. A common method used to resolve multipath and/or multiple signals is the use of high-gain narrow-beam mechanically rotating antennas [1], [2]. This method can resolve two or more signals provided they are separated by an angle greater than the beamwidth of the antenna. Thus, the number of signals that can be detected depends on the beamwidth of the antenna. Other techniques for minimizing the impact of multipath interference involve the implementation of frequency hopping, polarization agility, and space diversity. A previous paper discusses the use of four directional arrays to reduce the fading caused by multipath signals [3]. This technique uses four separate arrays, each with a directional pattern, and requires switching between these arrays.

The methods used to reduce the effects of multipath mentioned above have been successful, however, these systems are generally quite complex. Where possible, it is desirable to avoid mechanically rotating parts so that power consumption is kept to a minimum. The alternate methods have involved the use of phased arrays, allowing 360° rotation with no moving parts. This would seem an ideal solution, however, in order to track multipath signals, a full 360° sweep is required regularly so that the desired signal is distinguished from other spurious signals. This process is time consuming with both the phased array and the mechanically steerable arrays and with phased arrays it is also computationally intensive. Certain applications cannot afford the time spent in these cases, for example, with the TDMA modulation scheme used in GSM for mobile telephones, the minimum interval between frames is 4.038 ms.

II. ELECTRONICALLY STEERABLE ANTENNA ARRAY

The basic concept of an electrically steerable switched-parasitic antenna array has been presented previously [4], [5]. Near-resonance parasitic elements are used to create a directional electronically steerable antenna array. This technique can be applied to both wire antenna structures (e.g.,

0018-926X/98$10.00 © 1998 IEEE

FIGURE 2.1 An example of the first page of a refereed journal article. Note the following characteristics: the length of the title, the name and page details of the paper, the dates when the manuscript was received and revised, the index terms, the names and affiliations of the authors and the abstract. References are included in square brackets.

might be resubmitted in light of the corrections made in response to the initial review. The last date is the date of publication; recently this date might be the date when the article was published on the web and/or the date when the article appeared in print.

When submitting a paper for review, the authors are required to state that their article (in full or in part) has not been previously published and has not been submitted for publication elsewhere including another journal or conference.

In addition to publishing 'full papers', many journals will publish shorter works such as comments on papers, corrections to papers, short notes, technical notes and letters. These publications are all listed in the archival scientific and engineering literature. Readers of the full papers might sometimes find that their search reveals these additional short papers. In particular, factual and mathematical errors can be misleading and so searching the authors' names or the paper title can be very useful in finding more recent information pertaining to the field and any corrections to the original paper. These additional comments, after peer review, will be published, and will cite the original paper.

Some professional engineering associations also publish a magazine. These magazines can contain full journal-type articles and usually they are referenced in the same manner as journal papers. The articles are reviewed in a manner similar to journal papers, however, many items in the magazines will not be rigorously reviewed. Care must be taken when reading and citing magazine articles.

2.3.2 Short journal articles

Most journals allow the publication of short articles. These can be in the form of 'letters', 'short communications', 'comments', 'errata' and 'notes'. The reviewing procedure for most short articles is a quicker process. For some publications, only the editor or associate editor will review the submission.

Comments on a published full paper are submissions from other researchers who have read the article and have suggested errors, misinformation and/or a failure of the article to review the literature fully. The comment will be sent to the original authors for a reply. Usually both the comment and the reply will be published in the same issue of the journal.

Short communications are usually confined to the publication of new but minor discoveries, however, some short communications are used as a mechanism to gain speedy publication of new ideas. Both comments and short communications might have been assessed by an editor or associate editor of the journal only and a full review of the literature will not be included.

Some journals only publish letters. These journals require authors to submit short articles in which the number of words and figures is limited. The review process is still rigorous, but reviewers are asked only to provide the journal editor with a yes/no decision to publish. The authors receive no feedback on the article. If accepted it will be published as it was submitted. Because of the page limits, letters have limited explanations, descriptions and a smaller number of references. Letters are usually published much more rapidly than full papers.

2.3.3 Conference papers

Scientific and engineering conferences are meetings of engineering researchers with the aim of updating and reporting research developments not yet published. At these meetings researchers present research papers and discuss their latest findings, either through a formal presentation in a lecture room, or a poster presentation. Authors are usually required to submit a written paper to the conference technical committee. The paper is reviewed for relevance and correctness and the authors of the accepted papers are invited to make a presentation at the meeting. The papers are released to attendees at the conference as 'conference proceedings'. This is the printed record of the conference. After the conference is over, the papers may be made available to the wider scientific community via the web and accessible through the scientific web-based search engines. These papers usually contribute to the archival literature.

Conferences can be a preferred method for researchers to announce recent results as commonly the time scale between conference paper submission and publication is much shorter when compared to the journal review and publication process. Commonly conference papers are submitted six months before the conference and published on the first day of the conference. The rapid timescale of conferences means that the review process of papers is short or non-existent. If the conference includes a peer review process, the technical committee of the conference will assign the papers to one or two people on a panel of experts and the papers are usually reviewed on a pass/fail basis with only editorial changes allowed. Thus conference publications are usually not rigorously assessed and so should be regarded by novice researchers as important but subject to some uncertainty when compared to journal papers.

Figure 2.2 shows a typical first page of full paper a presented at a conference. The major features revealed on the front page include:

- The title of the paper (usually containing more than ten words);
- A list of the authors, their institutions and contact details;
- An abstract which outlines the work and major conclusions;
- A list of keywords relating to the paper.

Most conferences will publish their papers as 'proceedings' of the conference and will use an ISBN number (International Standard Book number) rather than an ISSN number. The volume number will be confined to the number of volumes published for the particular conference. In some cases, conference presenters are invited to expand their paper and submit it as a journal publication in a 'special issue' of the journal. In this case the normal reviewing process for journal papers is followed, but with strict time lines for submission and publication.

2.3.4 Books

There are three types of books commonly used by academics in their research:

Flexible, Light-Weight Antenna at 2.4GHz for Athlete Clothing

Amir Galehdar and David V. Thiel*
Centre for Wireless Monitoring & Applications
Griffith School of Engineering, Griffith University – Nathan
Queensland, 4111, Australia
E-mail: d.thiel@griffith.edu.au

Abstract

Linearly polarized rectangular patch antennas printed on light-weight cotton clothing are subject to both convex and concave bending. Changes in resonant frequency resulting from H plane bending are explained in terms of changes in effective length. Agreement between theory, practical and simulation results was observed. The resonant frequency changed by up to 8% for extreme bending although the bandwidth remains essentially unchanged at 4.5%. A 16.8% bandwidth was achieved using a double U-Slotted patch which minimized the problem.

Key words: wearable antennas, patch antenna, dual slots, curved patch.

I. Introduction

Athlete monitoring during sporting events and training is of significant interest to coaches and television broadcasters. Coaches seek biomechanical and physiological information to make tactical changes during the event. Television broadcasters seek to inform viewers as the status of the athlete. Wearable antennas for these applications must include sufficient antenna gain to minimize battery size and sufficient range to cover the field of play (usually not less than 100 m line-of-sight).

In this paper the effects of concave and convex bending in E and H plane on wearable antennas are reported. An increase in the bend angle for a concave H plane increased the resonant frequency. This is thought to result from a decrease in effective length. Convex H plane bending has the opposite effect. These results were verified by simulation, experiment results and an approximate theory. Concave E plane bending was measured and explained by previous researchers [1, 2]. An increase in bend angle decreased the resonant frequency.

This paper outlines the design of a 2.45GHz flexible patch antenna mounted on fabric having thickness 1.6 mm and relative permittivity $\varepsilon_r = 1.63$ (determined experimentally). The fabric adhesive used to fix the copper patch and ground plane increased the effective relative permittivity to 2.6. The basic design was built using flexible copper mesh as both the ground plane and the patch with fabric as the dielectric. Adhesive backed copper-coated non-woven nylon fabric is available from commercial suppliers for EMI suppression applications.

FIGURE 2.2 An example of a conference paper. Note that the paper title is shorter, the author's names and affiliations are given, the paper has an abstract and the page numbers are given. Commonly the name of the conference does not appear with each article. References are included in square brackets.

- Textbooks;
- Research books (monographs); and
- Reference books.

Textbooks are used in undergraduate and postgraduate courses and in training people to enter the engineering profession. These books usually have a very large circulation and often have repeated editions as the material is upgraded and errors are corrected. Commonly the titles of textbooks contain 2–4 words and are designed to be used for one or two courses/subjects in tertiary educational institutions such as universities. Often several textbooks will have the same name and so the author, publisher and date of publication must be included in references to the book. High quality textbooks are commonly re-released with corrections and additions as appropriate and so the edition number of the book is an important inclusion when citing the book in a research publication.

Textbooks usual cover standard experimental, theoretical and computational techniques used in the field. Most textbooks have detailed references and bibliographies. When writing a research paper, the author need not rewrite the well-accepted theory, but rather an appropriate textbook can be used as a reference and a source of equations, definitions and standard experimental methods.

Research books are written by experts in the research field. The target audience is the small, more specialised academic community and the books contain higher level information on a specific topic.

A third type of book is the reference book. These books commonly contain an alphabetically ordered index of terms. These books might include 'dictionary', 'encyclopaedia' 'reference' in the title.

Paper and electronic copies of books are subjected to professional examination between the time of writing and before publication. Commonly the material in a textbook is well established in the field and presents only that knowledge and understanding which reflects the current position.

The front matter in every book includes the book title, author and affiliation, publisher and place of publication, an ISBN number, the date of publication and the edition number. Most of this information must be used when referencing the book in a scientific paper.

2.3.5 Standards

The engineering profession usually provides practising engineers with standards. A standard is a document that defines a particular experimental technique or a requirement specification. Standards can also be used to define engineering terms so that the profession uses terms in a well-accepted and defined way. Novice researchers should be familiar with these terms and their precise definitions, and use them correctly when writing research papers.

Engineering standards are reviewed as technology develops, and new standards are written and approved by experienced members of the profession in the relevant discipline. Following a review, some changes will be made and it may be that some previously accepted terms are no longer acceptable. Such terms are referred to as being 'deprecated'. Such changes are made to resolve confusions and to add new terms as the technology develops.

Most standards are published using an ISBN number (i.e. a book reference number). While the names of the committee members who developed the standard are listed inside the standard, the reference will be to the professional society rather than any one individual or group of individuals.

The International Organization for Standards (ISO) and the International Electrotechnical Commission (IEC) maintain standards internationally; however, most countries have national standards authorities which are charged with the maintenance and implementation of national standards.

Example 2.4 **Examples of standards**

In the IEEE-SA Standards Definition Database [4] are definitions of electrical, electronic and software engineering terms.

In mechanical engineering, the mechanical properties of materials are measured using standard test procedures [5].

In geotechnical engineering there are standard methods for testing the shear strength of soils [6].

It is wise for a research team to use measurement techniques and numerical computational techniques which are defined by the most recent standard. Lesser known measurement techniques might lead to incorrect results which may compromise the research outputs.

Commonly the standards terms and techniques are reproduced in part in textbooks and so seeking the original standards documents (often at quite high cost) may not be necessary.

2.3.6 **Patents**

A patent is a document written to protect an idea for commercial advantage and exploitation.

Patents are written and granted on the basis of their originality (referred to as an 'inventive step'). The object of a patent is to protect an invention or innovation against commercial theft whereby one company makes a profit from an invention by another person or company. While each country maintains a patent assessment process and an inventory of patents granted, most patents are readily available using a web based search using keywords. Most patents will cite previous patents and other published literature.

Patents are country specific – that is, most nations manage their own patents through a patent examination procedure before the patent is granted. Figure 2.3 shows an example of the front page of a registered patent from the United States of America; however, the patent format is different for different countries. The

US006034638A

United States Patent [19]

Thiel et al.

[11] **Patent Number:** 6,034,638

[45] **Date of Patent:** Mar. 7, 2000

[54] **ANTENNAS FOR USE IN PORTABLE COMMUNICATIONS DEVICES**

[75] Inventors: **David V. Thiel**, Cornubia; **Steven G. O'Keefe**, Chambers Flat; **Jun W. Lu**, Wishart, all of Australia

[73] Assignee: **Griffith University**, Queensland, Australia

[21] Appl. No.: **08/557,031**

[22] PCT Filed: **May 20, 1994**

[86] PCT No.: **PCT/AU94/00261**

§ 371 Date: **Mar. 14, 1996**

§ 102(e) Date: **Mar. 14, 1996**

[87] PCT Pub. No.: **WO94/28595**

PCT Pub. Date: **Dec. 8, 1994**

[30] **Foreign Application Priority Data**

May 27, 1993 [AU] Australia .... PL 9043

[51] **Int. Cl.**[7] **H01Q 1/24**

[52] **U.S. Cl.** **343/702**; 343/815; 343/841; 343/873

[58] **Field of Search** .... 343/702, 785, 343/700 MS, 790, 815, 818, 841, 851, 833, 834, 876, 872, 873

[56] **References Cited**

U.S. PATENT DOCUMENTS

3,268,896	8/1966	Spitz	343/785
3,541,567	11/1970	Francis et al.	343/873
3,560,978	2/1971	Himmel et al.	343/833
3,725,938	4/1973	Black et al.	343/833
4,123,759	10/1978	Hines et al.	343/854
4,170,759	10/1979	Stimple et al.	343/876
4,356,492	10/1982	Kaloi	343/700 MS
4,367,474	1/1983	Schaubert et al.	343/700 MS
4,379,296	4/1983	Farrar et al.	343/700 MS
4,414,550	11/1983	Tresselt	343/700 MS
4,631,546	12/1986	Dumas et al.	343/833
4,700,197	10/1987	Milne	343/837
4,800,392	1/1989	Garay et al.	343/700 MS
5,075,691	12/1991	Garay et al.	343/830
5,243,358	9/1993	Sanford et al.	343/700 MS
5,338,896	8/1994	Danforth	343/702
5,373,304	12/1994	Nolan et al.	343/841
5,507,012	4/1996	Luxon et al.	343/841

FOREIGN PATENT DOCUMENTS

214806	8/1986	European Pat. Off. .
588271	9/1993	European Pat. Off. .
2216726	3/1989	United Kingdom .
2227370	11/1989	United Kingdom .

*Primary Examiner—*Don Wong
*Assistant Examiner—*Tan Ho
*Attorney, Agent, or Firm—*Jenkins & Wilson, P.A.

[57] **ABSTRACT**

In one embodiment, an antenna has four equally spaced monopole elements mounted in a symmetric array on the outer surface of a solid cylinder structure. The cylinder has a high dielectric constant, and extends from a conductive ground plane. The monopole elements can be switched by switching elements so that one or more is active, with the others acting as parasitic directors/reflectors being connected commonly to ground or left in an open circuit condition to be effectively transparent.

22 Claims, 9 Drawing Sheets

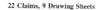

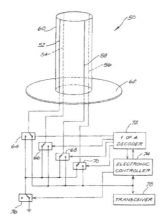

FIGURE 2.3 An example of a US patent. Note the inventors are listed together with their affiliations, the date of filing and the previous work (references cited) are listed. There are a number of codes for the sub-discipline of the invention (both national and international). The abstract summarises the invention.

International Patents Treaty between nations ensures that there is some uniformity between the national patent offices.

A patent is characterized by a patent number, the date when the patent was submitted, the authors of the patent and their affiliations, the sponsoring company, a brief summary and a series of six number codes which define the field in which the invention will find application. For example the first page of a USA patent (Figure 2.3) includes the 'Field of search' codes as 343/815 where 343 refers to a general subject area (antenna) and 815 is a more specific (subclass) of the area (with radio cabinet) [7]. The previous patents and papers in the area (called 'prior art') are also listed under 'References cited'. The Abstract briefly outlines the subject material of the patent. The novelty claims of the invention are listed numerically at the end of the patent together with a series of figures.

Figure 2.3 also includes the International Classification Code H01Q 1/24. Using the index [8] one can determine that H refers to electricity, H01 refers to basic electric elements and H01Q defines aerials.

Many scientific and engineering technologies are protected by patents and this can restrict their use in both the research and commercial environments. Using a patent for commercial benefit without the permission of the owners of the patent is regarded as an infringement and may cause legal action. Inventions (as defined by patents) can be purchased and sold and so the current owner of the intellectual property might not be listed on the patent. It is the inventors who must be listed if reference is made to a patent.

Patent applications filed for consideration are assessed by the national patent office of the country in which patent protection is sought, and will be granted only if the technology or methodology is an inventive step. The time delay between the original submission of a patent application and the registration of the patent can be up to 3 years.

The award of a patent does not imply that the technology functions in the manner described or that the method has advantages over existing technologies, despite any claims made in the patent document. Thus, when conducting a literature review, researchers must understand that while patents contain new ideas and novel applications of these ideas, they are not a reliable source of scientifically verified advances in the field.

2.3.7 **Theses**

Many tertiary education institutions require their undergraduates, master's degree and PhD candidates to submit a thesis as part of their final assessment. If the student is awarded the degree, then the thesis may be available on the internet and will be identified during electronic searching.

The status of these documents varies with institutions. Commonly most bachelor's and master's theses are marked on the basis of a pass/fail system and there is no revision of content in line with examiner's feedback. For this reason, theses can provide useful information but might not be a reliable source of competent, verified new knowledge. PhD theses are usually corrected in line with examiner's feedback and so constitute a more reliable source of information.

Most institutions require PhD candidates to publish their work in the international scientific literature and so preferred references are to the published papers rather than the thesis itself. It should be noted that all theses bear the name of the candidate only and will not include the names of other members of the research team. This is another reason to refer to the published papers from the thesis rather than the thesis itself.

Theses often carry useful information about experimental methods which are not always described in their entirety in journal or conference papers. In electrical engineering it is common to include circuit diagrams, in mechanical and structural engineering it is common to include mechanical drawings and in chemical engineering, comprehensive details of chemical processes. If used in further research, then these details should be referenced to the relevant thesis. Researchers need to remain cautious about the accuracy of the material in theses.

2.3.8 Trade magazine articles

Trade magazines contain large numbers of advertisements as well as some articles written by the editorial staff of the magazine covering the latest developments in a particular sub-discipline or field of research. The writer is likely to be a scientific reporter rather than an expert in the field and will be reporting information previously released by research institutes and companies as press statements and other forms of publicity. These articles are generally review articles covering recent journal papers and/or company releases about new products.

While such articles are generally informative, the articles are brief and lack scientific and engineering detail. These are not a particularly reliable source of new knowledge. They are secondary sources and so researchers should cite the original articles.

Figure 2.4 is an example of a trade magazine article. Readers will note that the title is relatively short, there is no abstract/summary, there are no keywords, and the author's affiliation and contact details are not given. These articles commonly have very few references.

2.3.9 Newspaper articles

Most general newspapers provide their readers with commentary on recent scientific and engineering innovations. As with trade magazines, the articles are quite short, lack detail and have a tendency to be sensationalist (to sell more papers). These articles do not contain a reflective summary on the research outcomes and so are of limited use as a research resource.

Figure 2.5 is an example of an article published in a national newspaper. Note that the title is short, the author is a journalist and not a researcher, and there are no references.

MEDICAL

New electrode designs improve cochlear implants

AUSTRALIAN academic and commercial researchers have developed designs for higher-performance electrodes which could substantially improve sound perception in the next generation of cochlear implants.

The work was carried out by University of Melbourne Research Fellow Dr Carrie Newbold, the HEARing Cooperative Research Centre (HEARing CRC) and Cochlear Limited.

The research, which has been ongoing since 2001, looks at less intrusive, slim electrode designs, the use of new biomaterials and manufacturing techniques to produce electrodes with higher capacity for information transmission.

The new designs are based on Dr Newbold's research into the

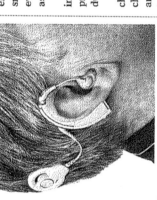

The next generation of cochlear implants could benefit from slimmer electrode designs.
Image credit: Cochlear.

interactions between the cochlear implant's electrodes and the tiny nerve cells of the cochlear.

The current design of the cochlear implant uses 22 individual electrodes. Because they are so small, their connecting wires are even thinner. This necessitates hand assembly, limiting production.

The delicate membranes and internal structures in the cochlea present an additional challenge during the implant process.

The new designs address these challenges by changing the physical characteristics of the electrode array and making it easier to surgically insert the device with minimum risk of damage.

The researchers are also looking at the potential application of new conductive polymers for improving hearing stimulation in the cochlear. These replacement plastics are more efficient with electricity, and are also less brittle. ∎

www.hearingcrc.org

FIGURE 2.4 An example of a complete article in a trade magazine. The title is relatively short and the name of the author is not included; only the name and brief contact details of the sponsoring organization are given. There are no references in the article (reproduced with permission, *Electronics News*, p. 4, 14 June 2013).

THE POWER WITHIN

A bioenergy harvest from the heart or inner ear could help power devices such as pacemakers in future

SHIRLEY WANG

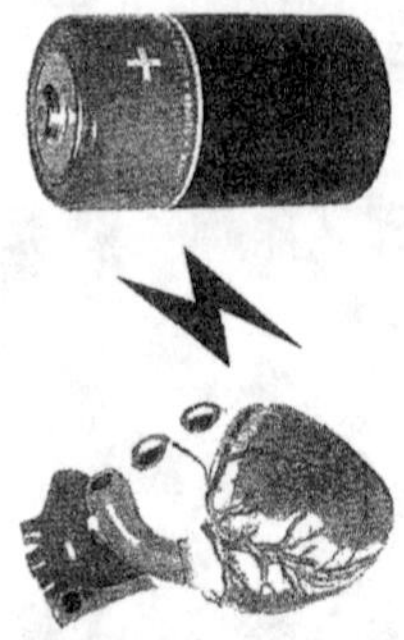

> Researchers compare the futuristic devices to solar-powered calculators

SCIENTISTS are studying how to tap the energy naturally created by people's bodies — such as heat, sound and movement — to power medical devices without the need to change batteries.

The development, still years from becoming a reality, could spare some of the millions of people with implanted devices like pacemakers from undergoing surgery to replace rundown batteries. Other products, including hearing aids, insulin pumps and pain-management devices, could be made to function without changing batteries, or at least sharply extend their power time. Harnessing the body's energy also could spur development of medical technologies that could potentially monitor the body's inner workings.

In an experiment this month, researchers from Massachusetts Institute of Technology and Harvard Medical School demonstrated in guinea pigs that it is possible to use energy produced by the inner ear, which is needed for hearing, to power tiny sensors without interfering with the animals' functioning. The researchers say the device, which resembles a semiconductor chip, is small enough to be implanted in the ear to monitor problems such as ear infections and hearing loss.

Scientists at other labs are trying to capture and convert energy from heartbeats, blood flow, lung contractions and arm and leg movements.

Researchers compare the futuristic devices to solar-powered calculators, which work as long as there is sunlight. Some experts expect the first medical devices that tap the body's energy — known as bioenergy harvesting — could be available within a decade.

Pacemaker batteries typically last for five to 10 years before they wear out and need to be replaced with a new battery using surgery. Currently, there is no way to recharge the pacemaker battery once it's depleted. In the future, the hope is the battery will be continually powered by the energy from heartbeats. The ultimate goal "is what I call the perpetual device", says Gene Frantz, an expert in power dissipation and an engineer at semiconductor design and manufacturer Texas Instruments in Dallas. "It doesn't need to be plugged in or operated with" an external battery, he says. Frantz studies innovation in integrated circuits, including work from many academics and companies researching bioenergy harvesting.

The task is challenging on several fronts. Many biological sources of energy tend to emit very low power and often inconsistently, so the energy needs to be captured and stored, such as with batteries, so it can be used by the device, say researchers. Also, energy storage units have to be limited in size if they're going to fit within the body.

An additional hurdle: the devices must not siphon off too much energy, so as to preserve normal body function, says Paul Kohl, a Georgia Tech professor and director of a centre that funds researchers, including the MIT group, with grants from the Semiconductor Research Corp, a technology research consortium.

In the work published earlier this month online in the journal Nature Biotechnology, Konstantina Stankovic, an ear surgeon, teamed up with MIT engineering professor Anantha Chandrakasan to figure out a way to tap a reservoir of energy produced by the inner ear to power a medical device. Scientists have been aware of this source of energy, which we need to use to hear properly, for about 60 years, but didn't know how to access it.

The researchers designed tiny electrode sensors that could be implanted in the inner ear and would draw just a small bit of current while maintaining normal hearing, says Chandrakasan, head of MIT's department of electrical engineering and computer science.

It was an engineering feat to design such an ultra-low-power device considering that the ear produces just 70 to 100 millivolts — not even enough juice to power a conventional electronic circuit like a sensor or Wi-Fi chip. A double-A battery generates more than 10 times that amount of voltage, says Chandrakasan.

The engineers found a way to kickstart a radio-frequency device wirelessly using an external power source initially, but running the device afterwards on the output of power generated by the inner ear. To reduce the need for power, they designed a device to wake up and run only when it was taking measurements from the ear.

Stankovic, an otologic surgeon at the Massachusetts Eye and Ear Infirmary, and a colleague implanted the electrodes into the inner-ear reservoirs of guinea pigs and connected them to the chip outside the animals. The chip, in turn, wirelessly powered a radio-wave device that was located a metre away and measured energy from the ear.

The scientists showed this biological battery was indeed able to run the radio device and detect the inner-ear environment for about five hours without compromising the guinea pigs' ability to hear. The device is small enough to implant into the middle ear with electrodes connected to the inner ear.

The most immediate application for the device is sensing the inner-ear environment and its vicinity, which can be important to monitor for infection in people with ear disease or a Cochlear implant, according to Stankovic.

Eventually, people with some hearing loss could be fitted with these devices to monitor hearing loss. This could help doctors understand why people lose hearing. The sensors could be used to measure treatment success with those receiving experimental therapies to regrow cells that restore hearing or to deliver drugs.

A University of Michigan team is working on harvesting energy from heartbeats to power a pacemaker. M. Amin Karami, an aerospace engineering research fellow, presented data at the American Heart Association annual conference last month showing an experimental energy harvester can convert the energy from the beating of a heart into enough usable electrical energy to, in theory, operate a pacemaker. The harvester hasn't yet been implanted in animals or humans yet.

Their team showed that the device could generate 18 times the amount of energy needed to power a pacemaker using a normal heartbeat, according to Karami. The group measured the vibrations emitted in the chests of pigs and sheep during surgeries, and then used those calculations to estimate how much electrical energy would be generated by the harvester.

"Our next step is to fabricate a complete self-powered pacemaker and examine it with animal tests," wrote Karami in an email.

At Virginia Polytechnic Institute and State University, electrical engineering professor Dong Ha and his team are in the early stages of studying how to employ body heat to power a circuit. Energy is generated by the body's adjustments to differences in temperature between the skin and the outside air, which can be harvested. If successful, such a device could be implanted just under the skin, or even worn in a jacket, to power a device, says Ha.

THE WALL STREET JOURNAL

FIGURE 2.5 An example of a newspaper article related to an engineering discovery. The name of the author is given and the names of various researchers and their institutions are mentioned in the article. The article is an overview of a number of new developments, and has been taken from another publication (*The Wall Street Journal*). This is clearly a secondary source of information. Readers of this article can find more information about the developments by searching the archival literature using the names and keywords (reproduced with permission, *The Australian*, p. 16, 3 December 2012).

2.3.10 **Infomercials**

This word describes video clips and written articles which are produced by a company with the aim of selling their expertise and products. The articles contain some factual information and specifications but little of the research approach used to develop the product (this would be of great benefit to competing companies). As the articles are not subjected to rigorous review by independent experts, the claims are unsubstantiated. These should not be used as reference material for researchers.

Infomercials can be found in newspapers, trade journals, web sites and might resemble a newspaper article or scientific article, but these are usually clearly labelled as not part of the normal technical content.

2.3.11 **Advertisements**

Advertisements on social media, television, magazines and newspapers contain little factual information and the claims are not supported by reliable evidence. They have not been subjected to independent review. For this reason researchers should not use advertisements as reference material.

2.3.12 **Wikipedia**

This is a web based resource (www.wikipedia.org) commonly found by web search engines. Wikipedia articles contain definitions of terms, history and scientific and engineering facts. The articles are written by the general public which includes experts. The articles are well referenced in the scientific and engineering literature and so the articles constitute a secondary source of information rather than a primary source. The articles can be changed at any time by any individual from around the world, and can be changed back in the same manner. There are adjudicators who monitor the activity on the Wikipedia pages and will restrict changes if they appear frivolous or biased.

Engineering and scientific papers should not include references to Wikipedia directly because it is subject to change. However, the original sources cited at the end of the article can be a valuable resource. While a researcher might cite a Wikipedia article to demonstrate changes or misunderstandings of some people, no other citation appears justifiable in the scientific press.

2.3.13 Web sites

Web search engines which are specifically set up to search the refereed scientific literature will yield web sites with scientific and engineering information. Novice researchers should take care not to use the material from web sites unless the following guidelines are applied:

- The authors and their affiliations are clearly found and can be considered reputable;
- The date of 'publication' on the web can be found.

Some universities and individual professors publish their lecture notes and laboratory notes on their web sites. These can prove to be a valuable resource for those not familiar with the field. In most cases, however, there is little new knowledge published on the web as researchers seek to gain credit for their work through refereed journal and book publications.

2.4 Measures of research impact

There are many refereed scientific and engineering journals and so an extremely large number of published articles. The number appears to increase exponentially. The principal readership of these journals is other researchers in closely related areas of research. While it is quite difficult to establish how many people read a particular paper (for on-line journal articles the number of downloads is one such measure), the reference in a paper to another published paper is called a citation or 'cite'. It can be argued that the inclusion of a reference to a paper in another paper is evidence that the cited paper formed the basis on which the new research was undertaken. For this reason, and because automated counting of citations is possible, one can review the number of citations of a published paper [8]. A paper might have a very large number of citations because:

- It is very innovative (e.g. through the inclusion of groundbreaking research);
- It is a review paper with many references (and so subsequent papers need not contain a comprehensive review of a large number of papers individually to satisfactorily demonstrate background knowledge in the field); or
- It might be incorrect (i.e. all of the citations discuss the errors in the paper).

Thus citation counts alone cannot be used as an assessment of the impact or value of the research outcomes reported in the paper. However, citation counts do constitute a measure of interest in a particular paper. It should be noted that most scientific papers have fewer than five citations. This indicates that while the journal editorial staff and the reviewers thought that the published paper has novelty and merit, the interest in most published papers is not high.

Clearly the older the paper, the longer the time available to readers, and the more citations that the paper might accumulate. Thus a paper published last month is unlikely to have any citations. Conversely, the older the paper, the less likely it is to be highly relevant as more innovations are published. There are some seminal papers in the field where time does not diminish the citation rate (i.e. the number of cites per year). This introduces the concept of additional measures of research impact, which include citations and the time over which these citations are made. The half life of a journal is an index which takes time and citations into account. These indices are sometimes used to evaluate the performance of researchers and the journals themselves. For example, the h-index for an individual researcher is the number of papers that have more than the same number of citations [9].

Example 2.5 h-index calculation

If Mary Jones has published 100 papers in her career and 12 of these papers have been cited 12 times or more, Mary has an h-index of 12.

It is clear that the number of citations depends on the database used. If the database includes journal papers, conference papers, books and magazine articles, then the citation count will be higher than an index where only journal citations are counted. The common computer data bases used in engineering are given in Table 2.2. Keyword searching, citation statistics and access to the paper abstracts and full text (if publicly accessible) are available from these databases.

While the h-index pertains to individual researchers, most journals are scored on their impact in the research field and the overall

TABLE 2.2 Databases used for citation related indices.

Database	Web address	Includes
Scopus [10]	www.scopus.com	journals, conferences
Google Scholar [11]	http://scholar.google.com.	journals, conferences, magazines, patents, books
Thompson Reuters, Web of Knowledge [12]	www.wokinfo.com	journals, conferences

body of knowledge. These measures are provided on a yearly basis. There are two accepted measures of the impact of a journal. The 'impact factor' is a yearly average of the number of citations per paper published in the journal [12]. Specifically it is the ratio of the number of articles published in the current year with citations to articles published in the previous two years divided by the total number of articles published in the previous two years.

Example 2.6 **Journal impact factor**

A journal volume 7 was published in 2010 and contained 100 articles. Volume 8 of the same journal was published in 2011 with 135 articles. That means 235 articles were published in these two years. The 2012 journal impact factor for this journal is the ratio of the total number of citations contained is all articels in all journals published 2012 to the number of published articels in these two years (i.e. 235 articles). If the number of citations in 2012 publications to these articles is 322 citations, then the 2012 impact factor of the journal is 322/235=1.37.

More recently 'Journal citation reports' (JSR) [12] have been introduced to normalise the impact factor data with an index which is typical of the research discipline. The source normalized impact per paper (SNIP) uses a weighting factor related to the subject field and the number of published papers in this field of research. The Scopus journal rank (SJR) [10] uses the SNIP index to rank journals in a specific discipline.

2.5 **Literature review**

It is mandatory that all research is grounded in the previous scientific and engineering knowledge. Thus, before commencing a research project and again immediately prior to the preparation of a research report (publication or presentation), the research team should undertake a review of the literature. Given the scholarly search engines available on the web, this is not an onerous obligation if the following simple task list is followed sequentially:

1   Key word searching;
2   Selection of relevant papers (partly influenced by publication date and citation number);
3   Review of paper abstracts for relevance;
4   Review of the complete papers of relevance;
5   Critical analysis of the results as they apply to the new research project.

In most journals, a reference list of fewer than ten papers is regarded as inadequate. Reviewers for journals are asked to assess the quality and relevance of the cited references. Papers that cite only the very recent work (particularly papers written by the authors themselves) is not highly regarded.

In reading a published article, the research team should note the following:

■   The relevance of the article to their research project;
■   The research methods described in the article;

- The conclusions reached at the end of the article; and
- The relationship of the article to other publications.

Example 2.7 Words for a review of a piece of literature

While not restricting freedom of expression, novice researchers should consider the following form of writing a review of a paper as a minimal effort:

Authors et al. (2018) investigated the effect of xxx on yyy. The research methods zzzz and aaa were used. They concluded that xxxxxxx.

The repeated letters *xxx, yyy, zzz, aaa* and *xxxxxxx* are the gaps which must contain the relevant descriptions of the work reported. Commonly the more highly relevant works will require more than this level of detail. The comments should be written in the past tense. A typical review of this type will have three or more sentences.

Following a number of such reviews, the literature review should conclude with a statement which groups the findings of others, critically analyses their strengths and weaknesses, and relates them to the current research project. It should be clear to the novice researcher that the current body of recently published literature is an excellent source of research project ideas.

Note that all literature reviews should be written using past tense as all the published work has been completed. The papers used in the literature review must appear in the reference list at the end of the publication, where there must be enough detail so that the reader can find the papers. Each reference must include a full list of authors, the title of the book or journal, the title of the paper, the volume number, page numbers and the date of publication. This can be done manually, but a number of software tools are available that can automatically collate references. Each journal has an individual reference style, so many of the citations must be reformatted to fit the required style.

Example 2.8 Referencing styles

Most engineering literature follows one of two possible referencing techniques:

The *numerical style* (e.g. Vancouver reference style and used by IEEE) uses references referred to by a number in square brackets, e.g. [1]. The numbers appear sequentially through the paper. That is [1] is mentioned in the text before [2], etc. The references are listed sequentially in the references section at the end of the paper. This is the method used in this book. If a reference is used more than once, then the same number is used and only one entry is given in the reference list.

The *author name style* (e.g. Harvard reference style) uses references in the main text as the first author's family name and the date of publication. Thus [1] would be written in the main text as (Newton, 1729). The reference section would then list the authors alphabetically. Repeat author names would be listed in order of the publication date. If the same author has two publications with the same date, then these are referred to as yeara and then yearb. Thus if there are two references with Newton as the first author dated 1729, they would be cited in the text and listed in the references as Newton 1729a and Newton 1729b.

A number of computer programs have been devised to assist in maintaining a reference list, and extracting references for inclusion in a publication in the style require by a particular journal. Such programs include EndNote, ProCite, RefMan, JabRef and BibTeX and many others. Care must be taken to read the

'Instructions for authors' for journals and conference papers, as there can be significant differences in the format required.

It is very important that every reference in the list of references is cited in the main text, and that every citation in the main text has a full and proper reference in the reference list.

A Bibliography is a list of papers consulted to develop an argument, but these papers are not referenced individually in the text. It is highly unlikely that a researcher in any engineering discipline will use a bibliography, but an instructor might provide a class with a list of relevant reading materials (i.e. a bibliography).

2.6 Keywords

Keywords are a critical part of the publication process if a publication is to be found by the internet search engines listed in Table 2.2. Most people are quite capable of conducting web based searches using keywords. However, some general comments are appropriate.

Keywords which are too general will define the field and little else. The result of the search will be an extremely large number of references. This is most appropriate for review articles. Keywords that are too specific will not be found by search engines, although the closest approximations will be given. That is, if five keywords are listed, then the search results must give papers in which three of the five words are relevant. Keywords should not include the commercial names of products (trade names including software products and hardware products). As most papers are confined to up to five keywords, care should be taken to use terms which are well accepted in the discipline. The use of acronyms (i.e. a list of letters which are the first letters from each word in a phrase) is discouraged as there are many meanings to most acronyms. There are websites which list the meanings of acronyms. If you visit one of these web sites, you will see the very long list of explanations for almost all acronyms.

Commonly the keywords should be different from the words used in the title as most search engines will include the title and the keywords in the same search. Note that most search engines will generate a list of references ordered using an algorithm based on the date of publication and the number of citations. That means the most recent papers with a high number of citations and most keywords will appear first on the list.

As publications take some time between the first submission and the final publication, and there is a time delay between the publication of an article and its appearance in the database, there is a substantial period of time between the completion of the research and the availability of the article. Given that conferences are commonly more recent when compared to journal publications, there is merit in including one or more of the authors in the keyword search. This will reveal any recent publications by those authors.

2.7 **Publication cost**

As the publication of research papers has become mandatory for must academics around the world, the number of papers and the number of journals and conferences continues to increase at a very rapid rate. This means that the services of editors, associate editors, technical committee members and reviewers are becoming more in demand. The rating of the journals using statistical measures (e.g. impact factor) provides significant competition in the search for authors and reviewers. Leading members of the research engineering profession have an obligation to review papers, just as their papers have been reviewed in the past by volunteers.

While in the past an editorial team was required to edit the papers, style the references and lay out the pages, much of this work is now done by the authors who must submit their paper within a designated template. Failure to do this results in automatic rejection. Following the acceptance of a paper, the editorial staff check for typographical and layout errors. All changes at this stage of production are referred to the authors for a final series of corrections.

There are number of financial models for covering the cost of publishing papers. As with any magazine or newspaper, there are publication costs which must be met or the journal will not survive in the world of business. In this section only the refereed international journal and conference papers will be discussed. Most other publications either charge the readership via a subscription (this is a newspaper cost model) or the costs are covered by paid advertisements. These are two commonly used financial models. There are others which are variations on these two approaches to meeting these publication costs.

The following subsections present a generalisation of the two common cost structure models used in scientific publications. Most journals and conferences have more than one funding source.

2.7.1 Publications and conferences hosted by professional societies

The archival literature of professional societies is funded partly by the membership fees of the society and the journal subscription is included in this membership fee for no additional cost to members. While in some cases the Editor-in-Chief receives some financial recompense for his/her effort and for his/her office including the editorial staff, the associate editors and the reviewers provide services on a voluntary basis. The cost of typesetting, printing and mailing (or maintaining the electronic membership list) is met by the professional society membership. Those non-members seeking access to a particular article published in the journal will be required to pay a fee to receive a copy. Most journals will have a page limit for papers. Papers that exceed this limit will require the authors of the article to pay for the costs of the extra pages. There may be a voluntary page charge for all authors, but the publication of a paper is not usually delayed or refused because of a refusal to pay this voluntary change.

The final printing is usually outsourced to a professional scientific publishing company. Their responsibility is to ensure timely, error free publication and the distribution of the final set of papers in each issue of the journal.

Professional societies host conferences on relevant topics. In this case, the authors of conference papers are not required to pay any publishing costs, but at least one of the authors is required to attend the conference to present the paper. There will be a conference registration fee which defrays publishing costs. With many conferences, a person who attends the conference is not allowed to present more than two papers. In this way, the cost of publishing the conference proceedings is covered by conference attendees. The conference Technical Program Committee and the reviewers of papers are not paid for their services.

2.7.2 Open access journals

Open access journals are made available to the public electronically (via the world-wide web) at no cost. Reader access is free. The funding for the editorial staff is covered by all authors whose papers are accepted for publication. Every paper incurs a publication charge. If this charge is not met then the paper will not be published.

2.8 Chapter summary

The refereed scientific literature is the strongest source of reliable scientific and engineering information. A literature review is an essential component in the planning and implementation of a research project. Copying information without acknowledging the source can provide researchers with major problems – both ethically and professionally. The refereed scientific literature is characterised by a long unique title, a list of the authors and their institutions and contact details, the dates of acceptance, keywords, and a thorough literature review.

When writing a literature review, an author should read the papers located via a keyword search. The contents of these papers, particularly the experimental method and the conclusions, are summarised in his/her own words. The summary must include the full reference to the paper, the nature of the investigation, the method(s) used and the significant results and conclusions from the work.

Web-based scholarly search engines can provide speedy access to previously-published works. Often these are ordered in terms of the publication date and the number of citations. There is a significant delay time between completion of the research work and publication in a journal. Novice researchers should search for more recent publications by the authors at conferences and in quick publication journals such as letters journals.

Exercises

2.1 Citation comparison: Consider a journal article in your discipline that was published approximately five years ago. Note the keywords and type them into one of the web-based academic search engines (e.g. googlescholar.com). Does the original article appear in the search results? How many citations does this article have? Have the same authors published further work in this field?

Compare the citations of this paper with those from the most highly cited paper in the search results? How many citations does this highly cited article have? If this paper was published before your original article, is it cited in your article? Do you think this high-cited paper should have been listed as a reference in your original article? Give reasons for your decision.

2.2 Copyright infringement: Consider a journal article published in the last two years. Cut and paste more than ten words from the introduction, and paste them into an academic search engine in quotation marks. Is your original article found? Are any other articles found? If so, you have discovered a possible copyright infringement. Novice researchers need to understand how easy it is to detect direct copying from previously published works. It must not be done, for ethical and legal reasons.

2.3 Mini literature review: Choose the keywords of your specialization and a particular project you wish to pursue in further research. Enter these keywords into a web-based academic search engine. Read the top five relevant articles that are freely available and write a literature review using the three-sentence structure outlined in Example 2.7. The review should include individual reviews of the five papers, a summary of the strengths and weaknesses of the papers reviewed, and the list of references using a consistent referencing style. The review should be approximately 300 words in length.

2.4 Copyright infringement: Investigate the legal definitions of copyright in your home country and other countries. Find the definition of 'infringement'. Using the two keywords ('patent' and 'infringement') and your engineering discipline keywords, find an example of an engineer or engineering company who was prosecuted for copyright offences.

2.5 Corrections to papers: Using your keywords, find an 'errata' or 'correction' to a published paper. Read the original paper and correction. Write a brief report on the error and explain why the error is sufficiently important to warrant the publication of the correction. (Often errors are discovered by other engineers working in the same field. These people might have tried to reproduce the published results without success. In this way, the error was detected.)

2.6 Definitions and standard research methods: Using a journal paper selected in your engineering discipline or interest, highlight at least three technical words and one experimental method. Find the definition of these terms and the experimental method either using a textbook, a published standard or a reference book.

2.7 Literature review quality: Using a journal paper selected in your engineering discipline of interest, write a 400 word evaluation of the quality of the literature review. In particular, review the quality and relevance of the cited papers, the comments made on those papers' contribution to the general field, and any omissions of papers which are of major importance in the field.

2.8 Journal impact factor: Using your keyword search in Exercise 2.1, note the impact factor for the first ten different journals (exclude patents and conference papers) returned by the search engine. Create a rank order of these journals. It can be argued that the publication of articles in those journals with very high impact factors will increase the likely number of citations of the article.

References

Keywords: engineering dictionary, definition, standards, citation metrics, impact factor, h-index

[1] Newton, I., *The Mathematical Principles of Natural Philosophy*, trans. A. Motte, London: Middle-Temple Gate, 1729.

[2] Maxwell, J.C., *A Treatise on Electricity and Magnetism*, Oxford: Clarendon Press, 1873.

[3] Martin, S.L. and O'Connor, A.K., *Basic Physics*, vol 3, Melbourne: Whitcombe & Tombs, 1950/60.

[4] IEEE-SA Standards Definitions Database, http://dictionary.ieee.org/dictionary_welcome.html accessed 29/11/2012.

[5] ISO 12135:2002: *Metallic materials. Unified method for the determination of quasi-static fracture toughness*, ISO, 2002.

[6] ISO/TS 17892–6:2004, *Geotechnical investigation and testing – Laboratory testing of soil – Part 6: Fall cone test*, 2004.

[7] United States Patent and Trademark Office, http://www.uspto.gov/web/patents/classification/uspcindex/indexa.htm, accessed 29/11/2012.

[8] World International Property Organization, *International patent classification codes*, http://web2.wipo.int/ipcpub/#lang=en&refresh=page, accessed 29/11/2012.

[9] *Publish or perish user's manual*, Tarma Software Research, available at http://www.harzing.com/pophelp/index.htm, accessed 29/11/2012.

[10] Scopus, www.scopus.com, accessed 1/10/2013.

[11] Google Scholar, www.googlescholar.com, accessed 1/10/2013.

[12] Thomson Reuters, thomsonreuters.com/web-of-science/, accessed 1/10/2013.

3

Developing a research plan

3.1 Research proposals

Like most engineering work, a research project should be structured and costed before it is commenced. This should ensure that the project plan is coherent and viable and acceptable to the project team and that all resources are available. In the tertiary setting, most universities require that students applying to undertake an internship, Master's degree or PhD degree must provide a one page outline of the topic as part of their application for admission. The project description is usually planned with one or more prospective supervisors, includes the background to the research, the contribution it will make, the resources required, the tools to be used and the likely outcomes.

For many researchers, a primary requirement might be to write research funding proposals. In this case the funding organization will require answers to specific questions. This process is very similar to a quotation (financial estimate) to undertake engineering work. As with engineering projects, the funding organization needs to be reassured that the research team has the necessary skills, a good track record, and that the results will be delivered 'on time and on budget'. Incomplete or inadequate research planning can lead to a failure to achieve the required outcomes, and the funding organization is unlikely to trust the researchers again. This chapter is aimed at providing guidance to the novice researcher so that the research plan is clear and comprehensive and the research team is presented in the best possible light, demonstrating the expertise necessary to achieve the planned outcomes.

There are many reference books and papers which assist novice researchers to plan research proposals [1–3].

3.2 Finding a suitable research question

One challenge in undertaking research is to find a place to start. Where can a research team find an opportunity to make a contribution to their engineering discipline and the world-wide body of knowledge? Some team members might have lots of ideas, but maybe some will have no concrete ideas. The research topic must lie within the range of skills and background knowledge of the team. In some cases the project will be limited to the tools and geographic locations that are accessible to the team.

Section 1.3 outlined the various types of research question. But writing the research question is very difficult in the absence of current knowledge of the research field. Usually the archival literature can provide the team with some guidance.

In the Chapter 2 exercises, a selection of keywords was used to find some recent journal papers in the field. A good scientific paper will include a discussion of the results and some suggestions for further work on some related topics. This is an excellent opportunity to develop a research question. But before prospective researchers get too excited about these ideas, they should investigate whether the research team that authored the paper might have continued with their research theme. This is not unusual as the lag time between paper submission and publication is at least six months. That means that other teams wishing to follow on with the published research must conduct a literature search using the authors' names and keywords. There is a possibility that their very latest results have been published in a recent conference. In addition is is possible that further papers have been submitted for publication and are currently being reviewed. Often a conference paper discovered in this way will report some preliminary findings of the next stage of the research. Perhaps the method was not sufficiently rigorous or the number of measurements was too small to be acceptable for a full journal paper, and so a conference paper was delivered to maintain the lead in this specific research theme.

Alternatively, the research team might have consisted of a PhD candidate and supervisors. In this case, it might be that the project has finished and the research theme has not been continued. If so, there is an opportunity for other researchers to follow up and extend this earlier work. An understanding of the make up of the research team which published the prior paper and members' individual roles can be useful in understanding the motivation, skills and roles of the team and their likely direction of further research.

It may be that the 'further work' section of the paper clearly indicates the nature of further work required and a further research question can be developed directly from the work. Alternatively, another research team might have access to a different approach which involves one or more of the following distinguishing features that allow additional developments in the same theme:

- Different equipment which allows improved accuracy and/or independent validation of the previous results;
- More powerful or more appropriate modelling tools which will verify the results with improved accuracy and allow extended modelling beyond the initial results that were published;
- A different researcher skill set which allows the previous results to be reviewed and analysed from a different perspective;
- A unique cross-disciplinary team with a variety of different skills which provide new applications in different research disciplines; and
- A novel idea to modify and improve the previous work.

In order to develop this idea, the team needs to study and understand the theory, experimental methods, data analysis, etc from the original paper.

If the outcomes of this preliminary investigation are positive and encouraging, then the new research extensions can be matched with one of the question words from Chapter 1:

How? Why? When? What?

With the addition of these words, a team will have a research question which has been derived from previously published work. The research question becomes the guide for the research plan. It has the additional advantage of reminding the research team of the ultimate goal of the research. Any additional research directions, tools, etc must be matched to the research question with the fundamental question:

Is this relevant to finding an answer to the research question?

3.3 The elements of a research proposal

The written documents used to describe research outcomes (see Chapter 7) all conform to a similar pattern. Not only do all research reports have this form, the research proposals themselves should have a similar structure. There are minor differences, but generally, writing a research project proposal is like starting to write a final research report. Figure 3.1 shows the basic structure of a research proposal. It includes a title, list of team members, a project summary and description, and budget. Note that when applying for a research grant, the research granting body might provide a template and/or structure which must be followed. Even when this is not the case, the standard structure described in this chapter is required. The essential elements remain the same.

The body that will approve the research project must be convinced that the suggested research topic lies within their areas of interest, that the research methods suggested in the proposal are sound and that there is a strong likelihood that the answer to the research question will be delivered with acceptable scientific rigour. Once the research proposal has been accepted and the necessary approvals given, the research team must maintain their focus on delivering the outcomes. The approval procedures for a research project are used to ensure that the central theme of the research is followed even if there is a temptation to stray from the original research question. A major change in the research direction must be sought and approved by the original approving body.

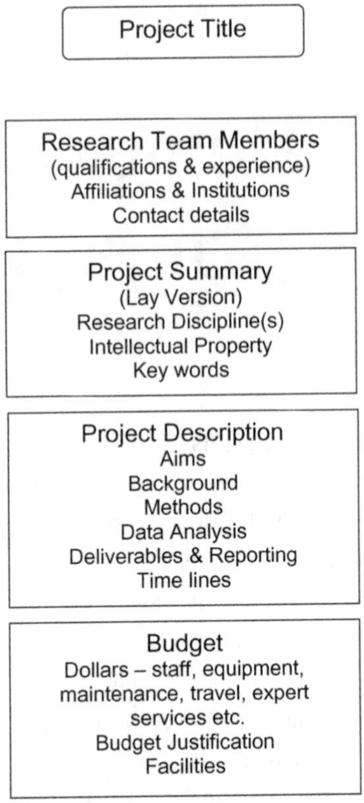

FIGURE 3.1 Generalized form of a research proposal.

Example 3.1 **Answering the research question**

A research team is funded to assess the environmental damage to local farms caused by a tailings pond from a nearby mine site. During their research the team discovers that one frog species living in the pond has exhibited a very unusual behaviour. One member of the team argues that the project team should conduct a study of this particular behaviour. If you are a team member, what is your reaction to this suggestion?

The elements of a research proposal are discussed in more detail below.

3.3.1 Project title

As is the case for the titles of scientific papers, the title of a research project should be clear and specific. More than ten words are recommended for the title. Acronyms should not be used as the assessment panel might not be familiar with their use. The title should indirectly indicate the engineering discipline of the research so that the readers can direct the project proposal to those who have relevant experience in the discipline.

3.3.2 Research team

It is rare for a project to involve only one researcher. The evidence for this is the very low number of single author research articles that are published. The research component of an internship, a master's degree programme or a PhD project normally requires the student and project supervisors to prepare a project outline before the student is enrolled/engaged. Thus the research team for one student will include at least one academic advisor. In addition the project is likely to need the support of other experts in specialised fields. For example, the team might include qualified personnel that will assist with the tests (setting up the equipment, maintaining the equipment, information technology personnel who are required to assist in the installation and maintenance of software and computing machine access, a physical chemist might be required to analyse soil samples, a statistician might be required to assist with the statistical analysis, a medical practitioner might be required for a biomedical project, a biomechanist might be required for a sports engineering project, etc). In some fields it is not uncommon to have more than 20 people as co-authors of a single paper. Engaging the services of additional experts outside the field of the principal researchers is one method of proving to the project plan assessors that a competent field of researchers has been assembled and is prepared to work together.

The project leader (often called the Chief Investigator or Principal Investigator) should take the lead in approaching people who will make a substantial contribution to the project. At the planning stages it is important to make it clear whether or not the individuals approached are likely to become co-authors on any publications arising from the research project. This preliminary agreement can reduce the possibility of a misunderstanding and disagreement when the time comes to write journal papers and to protect the intellectual property outcomes of the project using patents.

In some research fields the journal editors require a signed statement from all co-authors listed on the paper stating that they have made a substantial contribution to the research work, that they have read and perhaps written some sections of the paper, and that all named authors have given their approval for the paper to be submitted for publication in its current form. This is an ethical issue and an increasing number of journals is requiring such statements. One should not include the names of researchers who have not made a substantial contribution to the research project. Commonly and if appropriate, the technical help from laboratory assistants, the chemical analysis team, the computing support team, etc will be mentioned in an acknowledgements section of the paper rather than listed as co-authors.

The project leader is the person who will manage the project in all respects, who will be responsible for the expenditure of the budget, ensure that the milestones are met and the project outcomes are delivered. In higher education for a PhD thesis, the project leader is usually the research student, however, the thesis topic is commonly initiated by the primary academic supervisor and may be a subset of a wider ranging research project in the institution. Commonly students do not control the expenditure of research funds. This is usually the responsibility of the primary supervisor.

3.3.3 Project summary

While the project outline contains technical detail, the project team needs to convince the project assessment team that the project is worthy of funding. Often the assessment team consists of generalists who do not have specific detailed knowledge of the research topic and the analysis methods proposed in the project outline. The project team needs to provide a short, clear statement about how the project will benefit their organization and how it fits into the goals and objectives of the research field.

Example 3.2 A wider view of the research question

An antenna team aims to design new antennas that will improve the range of small wireless sensors used in a nursing home. The assessment team from the research granting authority might want to know the current and intended application of the wireless sensors and how an increased range will improve the lives of aged people living independently.

The project summary should also contain keywords so that the literature in the field can be thoroughly surveyed for competing technologies and research outcomes. A statement about who owns the intellectual property (IP) is usually required before the project starts. If the funding body or host institution sees the transition to market of the technology (products or services) as being commercially viable, these decisions must be agreed to in writing before the project starts. This usually requires a contract between the funding body, the university and the research student as part of a research team.

Example 3.3 Patent rights for students

A PhD candidate makes a significant advance in semiconductor technology while being funded by a multinational company and hosted by a state run university. Who owns the patent rights to the invention?

> ## Example 3.4 Intellectual property protection for students
>
> In most Australian universities the student owns the intellectual property (IP) developed as part of a research project. While the student can commercialize the intellectual property independent of the university, the contribution of the university supervisors, the university resources, and the funding company (who suggested the project) must be recognized. Most students prefer that the university or company undertakes the commercialization of the IP as the development and maintenance of an international patent is very expensive. The list of inventors on the patent includes the student and the university academics even when the company might own the patent. The organization which seeks to commercialize the IP will require written permission from each member of the invention team. This usually takes the form of an 'assignment of rights' contract.

3.3.4 Project outline

As is the case for all research reports (as described in Chapter 7), the project outline should include a number of sections. The research question is usually introduced early in the first section, which covers the aims and relevance of the project. One might then add a number of specific aims which clarify the methods of approach to be used.

The background section should contain a review of the literature and an argument that the proposed research will create new important, relevant knowledge. This analysis might include some of the initial work published by the research team. This can induce confidence in the assessment panel reviewing the research application. A theoretical understanding and the appropriate equations should be included where relevant, but only if the theory is relatively new or cannot be found in standard textbooks.

Example 3.5 Research question examples

Research question: What is the effect of local flooding on aquifer salinity?
Aims:

- to determine the permeability of the rock material between the surface and the aquifer;
- to determine the thickness of the rock material above the aquifer;
- to measure the aquifer salinity before, during and after flood events.

Research question: When will a MEMs based silicon micro-switch fail if it is exposed to the atmosphere?
Aims:

- to measure the surface contamination of silicon exposed to the atmosphere;
- to measure the mechanical properties of a silicon beam when subjected to continued use;
- to model the experimentally derived data using 3D multi-physics.

In the latter case, references to the basic theory are usually sufficient.

The research methods must suit the research question and the research aims in a clear and logical way. The type of measurements to be made, the potential difficulties with such measurements, the calibration procedures and the use of standard measurement techniques should be made clear. The research plan must include a description of the research methods to be used and the tools required and available for the project. The likely research outcomes that can be published must be described. The statistical support for the conclusions must be considered as part of the experimental method. Researchers should carefully consider how their results might be presented. In most cases of engineering research, the number of constraining variables greatly exceeds the parameters varied to influence the measurements. The presentation of multi-parameter analysis requires some thought, particularly if the research methodology cannot control many of the parameters influencing the measured results (e.g. natural variations in parameters such as temperature, humidity, sample purity, contamination effects, etc). These challenges are discussed in more detail in Section 4.3.

The data analysis section should outline the data processing required to remove interfering effects including calibration methods, and the type of statistical analysis that will be used to prove that the outcomes are valid 'beyond reasonable doubt'.

The project plan must clearly state a list of deliverables which will arise from the project. Deliverables are *things* which can be accessed by others such as reports, software code, presentations, images, etc. It is important that novice researchers understand that their own understanding of existing information is not a project deliverable. However, reports, presentations, patents, designs, objects, etc can be delivered to the granting authority as an indication of progress in the research project. New knowledge is only created when it is publicly available. In some cases, the research team might specify the conferences and journals where their results will be disseminated.

For commercially funded projects, the deliverables should be determined in consultation with the granting company/ organization. For 'public good' projects, it is expected that the outcomes of the research will be published in the refereed international scientific literature and accessible to all. There might also be a requirement for the researchers to participate in international conferences, to make presentations to the press and the general public, to assist in the writing of patents, and to engage in institutional advertising. These are the outcomes that various granting organizations might seek.

The timelines are particularly important in any planning process. This ensures that the granting organization can be kept up

Example 3.6 **Delivering deliverables**

An understanding of an FEM modelling computer program is NOT a deliverable. The functional characteristics of a device determined from an FEM structure modelled using the program IS a deliverable.

to date with research progress in a formal manner. For commercially funded research projects, there might be a requirement for monthly meetings to report progress. For government research projects the reporting might be required annually. Regardless of the formal reporting schedule, the research team should meet regularly to review progress against the projected timelines contained in the initial project proposal. While engineering companies commonly use Gantt charts to define the work breakdown, the time schedule and the deliverables, this level of complexity may not be needed for research project plans. Simple representations of the time lines using a spreadsheet or a simple text document can be used to indicate the work breakdown and the timing of deliverables (see Figure 3.2).

Concurrent engineering principles [4] should be used throughout the project and clearly indicated in the timelines given in the research proposal. Thus the timelines should indicate that a number of activities will be driven simultaneously. Figure 3.2 gives an example of these concurrent timelines.

At the start of the project, the literature review (an update on the literature review presented in the project outline) might run simultaneously with the ordering and installation of the test equipment, the acquisition of the samples to be tested and the organization of the data entry systems. In this way, the granting authority should be confident that even when there is a delay in one part of the project, other aspects can and will proceed.

Research timelines are difficult to predict. Most people find that research takes longer than any prediction as many unforeseen

Work	Q1	Q2	Q3	Q4	Q5	Q6	Q7	Q8	Q9	Q10	Q11	Q12
Literature review	X	X	X			X	X				X	X
Equipment arrival	X	X	◊									
Equipment calibration			X	X								
Sample preparation		X	X	X	X	X						
Data acquisition				X	X	X			X	X	◊	
Statistical analysis						X	X					
Numerical modelling	X	X	X			X	X	X	◊			
Report 1 submission		X	◊									
Report 2 submission							X	◊				
Theoretical development	X	X						X	X	X	◊	
Model optimization								X	X	X	◊	
Final report submission			X	X					X	X	X	◊

FIGURE 3.2 Simple representation of timelines for a project. The crosses represent major time requirements. Q refers to one quarter of one year. Only the major deliverables ◊ should be included. The work titles must be specific (not generic as presented in this illustration).

Example 3.7 **Research timelines**

In a research project, there is a long and unexpected delay in the delivery of a test instrument. The research team should have ensured that there are other allocated tasks to perform. These tasks should be clearly defined in the timelines of the project.

circumstances can occur. Some tasks can get delayed because of poor delivery times, bad weather, absence of members of the research team due to illness, resignations, vacations, etc. The break-through events occur somewhat randomly, even for experienced researchers.

It is common for research teams to undertake some preliminary investigations before submitting a research proposal to ensure that the project method has a very strong likelihood of success. Such initial investigations can be reported in the research plan to add strength to the proposal. In this way some of the research tools can be validated before the official start date of the project.

3.3.5 **The budget**

With the timelines defined in Section 3.3.4, it is possible to estimate the budget required to undertake the project. The budget
will require cash and 'in kind' support. The 'in kind' support is
a list of resources available to the project which do not require
direct financial support. For example, the host institution might
continue to pay the salaries of existing staff who work on the
project, they might provide facilities (space, electricity, internet
access, computing support, etc) without the need for additional
funding. In some cases the host institution will seek a contribution to these costs calculated as a percentage of the total cash cost.
The research team must engage in detailed planning to ensure all
resources, including human resources, are available.

Industry funded research projects usually operate under different funding rules compared to government research grants and
not-for-profit funded research. The financial arrangements will
be project specific and related to the 'rewards' structure of the
research. For example, if the research team can benefit finan-
cially from the research outcomes through the commercialization
of patents, etc, they would be expected to share in some of the
financial risk (i.e. the host institution is expected to provide some
funds for the project). If the research team does not benefit in this
way, then the sponsoring company can be expected to take all of
the financial risk.

The value proposition: When approaching an organization with
a research proposal, whether seeking funding or not, it is important to know and understand what is important to that organization. For example, the proposal should seek to address one or
more of the basic goals of the organization. A simple way to do
this is to imagine that the research team is making the funding
decisions in response to the project proposal. Given the organization's current financial situation and obligation to stakeholders
(e.g. the voting public, the shareholders, the executive council,
the students who participate, etc), would the successful completion of the project fulfil the needs of the stakeholders and generate

Example 3.8 **Funding model for industry research**

A university research team estimates that the cost of an industry funded project is $100K. If the University is allowed to publish the research outcomes or gains a share in the IP developed during the project, then the commercial enterprise and the university become equal partners in the research. Should the commercial partner wish to keep the research outcomes as commercial-in-confidence, then the university will charge $200K for the project. The additional $100K, over and above the project costs will be used by the university to fund other research projects not related to this project. In this way, the university researchers contribute to the university's performance indicators (e.g. published research outcomes). In addition the university research team, often funded by public money, does not undercut other research and development companies who need to make a profit to remain viable.

the publicity sought from the award of the research contract? If the research team can argue that the suggested research project will add value and enhance the reputation of the granting organization, particularly in the light of the time and money spent supporting the project, then the manner in which the proposal is developed should reflect the objectives of the organization. Of course, as practising engineers, those writing the proposal must ensure that the project will also be of benefit to society in some way. This remains an ethical obligation.

Table 3.1 lists the common expenditure categories and the possible division between research grant funded expenditure and the in-kind support for the project. Some research funding organizations limit their cash commitments to a percentage of the total research costs. Thus, the dollar values listed under research grant and in-kind should be equal if the maximum support is for 50% of the total project costs. If there is more than one participant institution, then an additional in-kind column

TABLE 3.1 Typical budget categories for a research project. Note that the items listed must be specific to the project. The items with a cross should be specified as numerical values in the appropriate currency (e.g. dollars).

	Project funded	In-kind contribution
Personnel		
Senior researcher 20%		X
Student researcher 100%		X
Electronic engineer 50% (circuit design)	X	
Research assistant 50% (data collection and analysis)	X	
Equipment		
Tensiometer	X	
Goniometer	X	
Network analyser		X
Maintenance		
Cleaning chemicals	X	
Test leads and adhesives	X	
Travel & conference		
Car hire (4 weeks)	X	
Flights (Sydney–Paris return) x 2	X	
Conference registration x 2	X	
Organization overheads		
	X	
Project total	X	X

should be included to list in-kind contributions from the second institution.

Every item in the budget must be justified with appropriate argument. The personnel list must have a clearly defined role for every researcher listed and funded in the project and this must be included in the budget justification. That means that the budget must quantify both the in-kind contributions and the personnel requirements sought from the funding organization. For those

Example 3.9 **Research contracts**

In some government sponsored research and development grant making schemes, it is mandatory that a commercial partner and a university sign an agreement to receive the funds. The research proposal will require a three-way contract between the government research granting authority, the commercial enterprise and the university research team. The budget will reflect the contributions from the commercial partner and the university in addition to the cash sought from the government grant in a king authority.

additional human resources not listed by name as researchers and not named specifically in the application for funding, a description of the skills and qualifications is required. This enables the researchers and the funding organization to clearly match the funds sought to the type of skills sought and this directly relates to the salary level defined in the budget. There might be a 'maintenance' cost included in the budget for advertising for the persons still to be appointed or this item in the budget might be added to the salary cost estimates.

The justification of the budget must match the research methods and analysis outlined in the research proposal. This is not a trivial task. It requires some thought and significant attention to detail.

It is necessary to state that the research equipment and research space not listed in the budget are available to the research team either as part of the budget or elsewhere in the grant application. The funding authority might wish to see a signed statement from an independent research facility demonstrating that they are happy with, and capable of, conducting the work to the standard required. Of course, if there is a cost involved, then this should appear in the budget.

The budget is a very important aspect of the research proposal. The research plan must be a cost effective strategy for the interested partners. An inflated, unjustifed budget is likely to result in no funding or reduced funding of the project. An underestimate of the research costs may result in funding, but a failure to deliver the research outcomes as stated in the research proposal.

Conflict of interest: When a research team receives funding from a commercial partner, some ethical issues might arise. In particular, researchers must ensure that the results are presented in an unbiased manner and that all conflicts of interest are clearly stated in any publications, presentations and reports. This can be done in the acknowledgements section of the paper.

Example 3.10 **Conflict of interest**

In some cases the conflict of interest is clearly evident. For example:

Research into the health effects of tobacco smoke is seen to be biased if funded by a tobacco corporation.

Research into the effects of gambling on society might be seen to be biased if funded by a casino.

An investigation into a bridge collapse is seen to be biased if conducted by the company that constructed the bridge.

An assessment of the health effects of medical implants is seen to be biased if funded by the manufacturer of the implants.

An assessment of the effects of deforestation on the conductivity of water in streams and aquifers funded by a timber mill will be seen as biased.

If funding has been received from an interested party, then the source of funding must be clearly stated in all reports and papers that arise from the research.

Potential conflicts of interest are often (but not always) avoided if the research team receives funding from a government department.

3.4 **Design for outcomes**

The research plan must be designed to give definitive answers to the research question. This must involve statistical support for the results. Multi-parameter investigations involve a number of variables which might or might not be controlled. This depends on the nature of the investigation. Thus if the researchers measure one parameter (the dependent variable) as a response to changes in another parameter (the independent variable) the consistency of the data is a measure of the strength of the data. This can be simply observed by looking at the scatter in the data points about the trend line, but a measure of this scatter is very important in establishing the accuracy of the trend line. While Chapter 4 provides additional details about various statistical analysis options, this section provides a short review of data presentation techniques and the type of results that will be provided to the funding organization or commercial partner as clear outcomes and proof of the validity of the research project.

Almost all scientific and engineering papers use figures to represent the experimental method as a block diagram and to summarise the results using graphs. When preparing a research plan, the novice researcher must consider carefully how the results can be presented in the most unambiguous manner. That is, a reader skimming the paper is likely to be attracted to the graphs which present the outputs of the research. These graphs must be self-sufficient; that is, the reader should be able to gain a level of understanding from the graphs and the associated captions without referring to more detailed explanations given in the text. Thus, a novice researcher writing a research proposal should consider how the readers might gain an appreciation of how the final results will be presented and what statistical analysis methods will be used to support any conclusions. There are many guides available to assist in this task [5–7].

Appendices A and B show some basic graphical techniques available using the software Matlab and MS Excel.

3.4.1 One-dimensional data

Some research projects seek to gain one piece of information only – a single number. In this book, we will refer to this as one-dimensional data.

Example 3.11 One-dimensional research questions

What is the conductivity of silicon carbide at 500 °C?
What is the rate of compaction of concrete piers under a multistorey building?
What is the optical reflection coefficient of soil?
What is the mean size of the nanoparticles?

The members of a research team must be able to convince the sponsors, scientific and engineering peers, and the world at large, that the number resulting from the research investigation is correct.

The presentation of one-dimensional scalar data is relatively simple. A single number with an error marking the upper and lower bounds constitutes a one-dimensional measurement. The same measurement of a number of samples represents an array of one-dimensional numbers. These data are commonly represented by a histogram, where the range of the measured values is

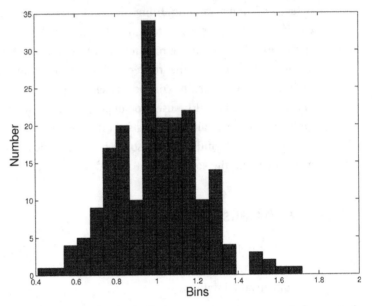

FIGURE 3.3 Two hundred voltage measurements were made from a strain gauge attached to a static beam. The measurements are plotted as a 20 bin histogram to illustrate the variation in the measurements. The bins are ranges of the measured voltage.

displayed (see Figure 3.3). The mean value of the data set can be presented as a one-dimensional result of the investigation. The standard deviation from the mean value is a measure of the variations in the measurements.

If the measurements are repeated, what might not be evident in the histogram is a small change in the mean value over time. Perhaps there is some time influence on the mean value? The simplest method to investigate this possible effect is to plot the data in the time sequence of measurements. Perhaps data that initially looked to be one-dimensional, might in fact be two-dimensional. The initial hypothesis as defined by the research question was that the value to be determined was a constant. The null hypothesis is that there is a variation with time. The probability of the time dependent relationship can be calculated using two-dimensional analysis techniques.

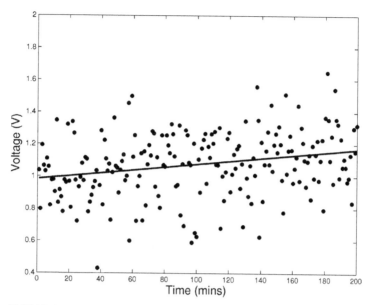

FIGURE 3.4 Scatter plot of the measured voltage as a function of time. The linear fit (continuous straight line) has a slightly positive slope, which indicates that the voltage might depend on the time elapsed.

3.4.2 **Two-dimensional data**

If a variable is measured as a function of time (for example), then a two-dimensional data set is generated by measurements. Commonly this will be represented by a line graph or scatter plot (see Figure 3.4). Note that if the data represent a continuous function a line should be used, but if the points are experimental measurements the individual points should be plotted without connecting lines. A theoretical line fit can be used to illustrate the theoretical relationship. This will be discussed further in Chapter 4.

Returning to the null hypothesis outlined in Section 3.4.1, a correlation analysis can be used to see if there is a change in the variable with time. A visual inspection of the graph might be inconclusive, however, the probability of a relationship existing must be estimated using standard statistical methods. These methods are discussed in more detail in Chapter 4.

Now that the one-dimensional problem has developed into a two-dimensional problem, so also the possibility of a third parameter might influence the dependent variable. Consider the possibility that temperature might affect the results. In making the measurements, assume that the temperature and the time of measurement were recorded. The research team is now in a position to evaluate the impact of temperature on the measurement. This leads us to a three-dimensional analysis.

3.4.3   Three-dimensional data

If a third parameter is recorded during the measurements of the original parameter, scatter plots on the same axes can still be used. It is quite simple to distinguish some points on the basis of this additional parameter. The data are now three-dimensional, but a two-dimensional representation can still be used. Assuming the additional parameter measured was temperature, Figure 3.5 includes circled points where the temperature was relatively high when compared to the total data set. It is possible to treat the high temperature values independently of the total population to decide statistically if there is a significant difference.

If the third parameter is controlled, multiple data sets can be plotted on a two-dimensional graph. There are a number of methods used to represent these data. Examples include the use of a contour plot, a surface plot, and a grey-scale image plot. As the data might not be sampled evenly, it is necessary to interpolate between points to create an array of regularly spaced points from which a three-dimensional image can be generated. In the example presented here, the temperature trend is not obvious from the scatter of the voltage data and so 3D presentations are not likely to reveal trends evident by eye. More powerful statistical techniques are required. These are discussed in Chapter 4.

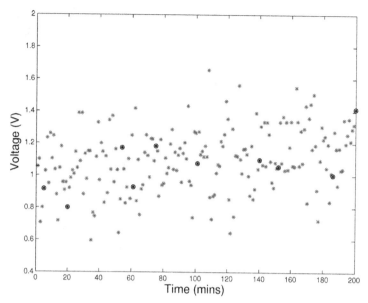

FIGURE 3.5 Scatter plot of the measured voltage as a function of time. Those values with a temperature in the highest 10 percentile have been circled. By applying an appropriate statistical technique it is possible to determine if the temperature effect is significant.

3.4.4 N-dimensional data

If more than three parameters are measured, a clear graphical representation of all of the data is not possible. Some software packages that generate three-dimensional data use transparent contour surfaces, a series of cross-section slices stacked side-by-side, and dynamic control over the viewing position outside or inside the model. The most common representation of these forms of data is through the use of a two-dimensional cross-sectional image with the 'cut' positioned through the features of interest. The research team has the challenging task of representing multidimensional data in reports in such a way that readers can be convinced of the authenticity of their conclusions. Tables of statistical parameters are one preferred method of doing this.

Example 3.12 **Research planning**

The research plan is to determine the size of a pollution plume from an exhaust fan. Researchers must address the following questions:

How many sampling points are required to define the plume?

How can the effects of temperature, wind speed and direction, seasonal effects, time dependent measurements be taken into account in a rigorous manner?

How can the final results be presented?

These questions all represent research challenges.

3.5 The research tools

In every research project, the challenge is to validate the results using independent means. The research plan must clearly outline the techniques which will be used to undertake this validation. The strongest case can be made if the experimental or numerical results can be validated against a theoretical model. If two very different and independent methods yield results which are statistically identical, then it is highly likely that both tools used in the research are valid. There still remains a small probability that both techniques yield results with the same bias; that is, both the results gained from the research project and the published results are incorrect. This might occur if the same error is common to both methods of analysis.

Example 3.13 **Systematic uncertainty**

The research team uses a published method of measurement. The conclusions agree with previously published work. This might be because there is a systematic error in the measurement technique which has not been identified. The net result is that both the published paper and the new results are incorrect. What methods are available to ensure that such problems are eliminated?

All research projects must be conducted in a manner in which workplace health and safety and quality assurance are addressed. These issues are discussed briefly in Chapter 8.

3.5.1 Experimental measurements

In most cases, standard measurement tools provide the best method of gaining reliable measurements. If alternative measurement techniques are used then the researchers must verify their tools through a calibration procedure. Researchers must plan ahead to ensure that they have access to the tools at the time they are required and that the tools are properly calibrated.

With most experimental systems, calibration is a well-defined procedure. Examples of calibration procedures include using provided standards (standard solutions in chemical analyses, standard loads for electric and microwave power level measurements, etc), the use of materials with properties that are well known, and 'null' calibration techniques when the instrument is run with a blank sample (e.g. air or pure water). Additional verification is possible by making measurements on materials and systems where the results have been previously published in the open refereed literature, but the source of such materials must be verified.

Example 3.14 Calibration in experiments

A new microwave vector network analyser is delivered with a set of standard microwave loads for calibration purposes. Measurements which are made without the standard calibration procedures will be very error-prone. When measuring the intrinsic impedance of a material half-space at microwave frequencies, the probe should be calibrated in air and against a copper sheet (a near perfect electric conductor at microwave frequencies). The calibration procedure should eliminate the effect of fringing fields, cables and connectors.

A UV-visible spectrophotometer uses a split beam light path. One beam is directed through a high quality, pure, solvent as a reference and the other passes through the sample under test. In this way the effects of temperature, frequency variations and the optical properties of the containers (cuvettes) are removed from the optical absorption measurements.

Many countries have measurement accreditation authorities. For example in Australia the National Association of Testing Authorities Australia (NATA) [8] is the recognised authority for validation of the equipment and measurement practices of organizations which undertake measurements on a commercial basis. This organization covers most engineering and scientific test requirements. The International Organization for Standardization (ISO) [9] provides documents which describe standard measurement techniques and many national government bodies and international professional societies develop and publish standard measurement techniques and guides to encourage best practice. Reference to and the use of these techniques adds strength to the research proposal and the research outcomes.

3.5.2 Numerical modelling

A plethora of computing codes is now available to model various physical effects [10, 11]. For example, codes in electromagnetics, computational fluid dynamics, and mechanical deformation (stress and strain) are now routinely used in predicting or verifying experimental measurements. Computing codes which are based on the solution of differential equations require a solution space which has finite dimensions (e.g. finite element and finite difference methods). Mixed codes (integro-differential codes) are also available. In most cases, the solution space is defined by a discretized area (2D pixels) or volume (3D voxels) representation of the object and its surrounding media subject to various external inputs.

For differential codes the media surrounding the object must be included in the model and the boundaries of the solution space must be sufficiently far away from the object to ensure that the boundaries of the solution space do not affect the area/volume of interest. This has a significant effect on accuracy. For integral codes, there is no requirement to define the boundaries to the solution space.

Problems which involve more than one set of physical equations are referred to as 'multi-physics' problems, and the various discretized sets of integral and differential equations are combined to solve problems which cover more than one computational discipline.

Example 3.15 Multi-physics computational methods

The dynamic response of a microelectromechanical (MEMs) lever switch requires the solution of mechanical and electromagnetic equations. As the switch moves, the electromagnetic environment changes and the restoring force on the lever changes simultaneously. The approximate solution to this problem requires a time stepped re-evaluation of the next mechanical situation as the lever moves under the influence of the force. This requires repeated solutions of similar structures which can be performed efficiently using small perturbations in the total structure. The mechanical model and the time variation are formed using discrete steps.

With software, it is necessary to ensure that the model is valid by computing the solutions of standard models and comparing the results with theoretical calculations. This can identify errors which are difficult to see using an image of the model. Common errors include imperfect continuity between materials and boundaries and material properties. The solution of simple models should match the analytical solutions found in textbooks.

There are also limits to the discretization. If the segments in the model are too small, then the program can experience rounding errors. If the segments in the model are too large, then the true representation of the continuous media is lost. A material boundary can be defined between two adjacent nodes, so that the distance between the nodes is the limit of the spatial resolution. It is good practice to change the mesh size in the same model and rerun the calculations. If the solutions are almost identical, then it is highly likely that the solution is valid.

3.5.3 Theoretical derivations and calculations

When developing a mathematical argument based on a set of theoretical equations, it is necessary to verify the solution. This can be achieved in a number of ways. For example, members of a research team could check their results as follows:

- Substitute their final equation back into the original equation to ensure that this provides a valid result;
- Check that the solution lies within the expected physical limits;
- Check the units of the result using dimensional analysis;
- Check the validity of the solution using mathematical programs (e.g. Matlab provides symbolic methods of solution for differential and integration equations);
- Substitute typical values into the equation to ensure that the result is reasonable (i.e. within expectation);
- Verify the boundary conditions for the independent parameters;
- Make very slight changes to the material parameters used in the model to ensure that the results do not change significantly.

In developing a research plan, these methods of code validation need to be outlined as a method of verification of the results.

3.5.4 Curve matching

Several methods are available to verify that experimental data follow a particular scientific mathematical equation. One technique is to apply the independent variables to the equation and calculate the dependent variable. A plot of the calculated dependent variable versus the measured dependent variable should be a straight line if the equation holds true for the data. A linear regression analysis of this plot will give statistical significance to the result. A visual inspection of the line will suggest the level of random error and data ranges where the relation fails.

Unfortunately, systematic errors might result in this method of analysis being inconclusive. Such systematic errors include the possibility of an offset angle in a trigonometic function (e.g. a sinusoidal or co-sinusoidal dependency). It becomes necessary to estimate the value of this offset so that the curve is matched with least possible error.

Consider the case of a polarized wave incident on a polarizing window. The radiation intensity detected on the other side of the polarizer should ideally be given by the equation

$$I = I_0 \sin^2 \theta. \qquad (3.1)$$

If the major axis of the polarized signal is not aligned with the angle $\theta = 90°$, and the polarizer is not perfect, this equation must be written with some additional, unknown variables:

$$I = I_0 \sin^2 (\theta + \theta_{\text{off}}) + I_1, \qquad (3.2)$$

where the offset angle θ_{off} and the crossed polarized component I_1 must be determined using an optimization algorithm. Figure 3.6a shows the raw data and Figure 3.6b shows the raw data fitted using Equation (3.1). An inspection by eye suggests that the data are not strongly linear although the correlation coefficient is 0.9841. There is clearly a systematic error as the deviation from a straight line is maximum in the centre of the plot. Figure 3.6c shows how the correlation coefficient varies as the offset angle is systematically changed in the equation. The best fit should occur when the linear correlation coefficient r is maximum (approximately equal to unity). Figure 3.6d shows the raw data fitted to Equation (3.2) using the best offset angle selected from the correlation coefficient curve.

A relatively simple technique of deducing the unknown values is to strategically estimate the values until the root mean squared error (RMS) is minimized. In this case the linear correlation coefficient r will be maximized. This is discussed in more detail in Chapters 4 and 5.

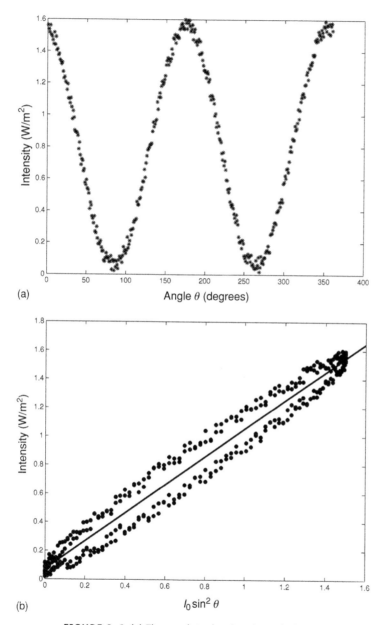

(a) Angle θ (degrees)

(b) $I_0 \sin^2 \theta$

FIGURE 3.6 (a) The raw data showing the variation in intensity with angle of the polarizer. (b) The raw data plotted using Eq. (3.1); the best-fit line is also shown. $I_0 = 1.55$ W/m^2. Clearly there is a systematic difference. (c) The variation in the linear correlation coefficient plotted as a function of the offset angle. (d) The raw data plotted using Eq. (3.2) with the best value of the offset angle. This occurs when the correlation coefficient is maximum.

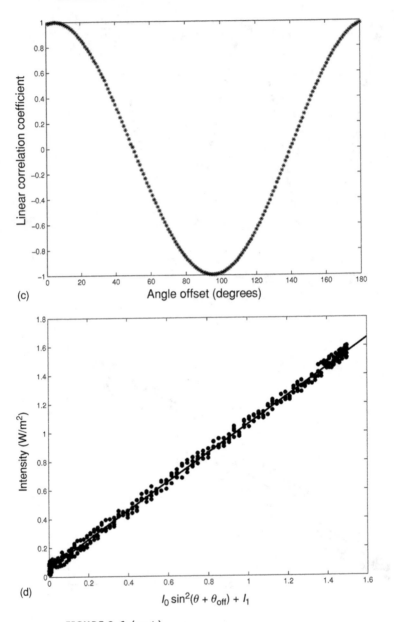

(c)

(d)

FIGURE 3.6 (*cont.*)

This example is relatively simple in that only three unknowns must be calculated. Chapters 4, 5 and 6 explain this process in some detail using a large number of unknowns which can be deduced using this RMS error minimization technique.

3.6 Chapter summary

A well prepared research plan will articulate the nature of the research question to be solved, the requirements needed to acquire the data to solve the problem, the expertise of the research team, the time and funding required to conduct the research work, the method of transmitting the results to all stakeholders including the engineering research community, the manner in which the results will be presented and the validation techniques which will be used.

Research results must be verified using two or more different techniques. Limiting cases and simple structures can be used to verify the research methods.

During the research planning stage, the research team should clearly articulate how the results will be presented. The method of presentation must be sufficiently clear and logical so that those reading the final report will arrive at the same conclusions as the research team.

In bidding for funding and other support, the research plan must clearly outline how the information will be presented and the benefits of the research findings to the funding organization and the community.

Exercises

3.1 Develop a new research proposal from a published paper: From a selected published journal paper, read the paper. In particular read the discussion and conclusions section and find suggestions for further work. Apply one of the question words (How?, Why?, What?, When?) and write one or more research questions arising from this paper. This can be used as a guide to help you develop your own research project proposal.

3.2 Outline the research proposal from an existing paper: Following on from Exercise 3.1, write down a project proposal which matches the theme of the paper and its outcomes. In particular, write down the research question you think was addressed, discover the skills of the research team (i.e. what contributions did each of them make?), and the elements of a research proposal as outlined in Figure 3.1.

3.3 Writing a new research proposal: Following on from your investigations in Exercises 3.1 and 3.2, write a complete research proposal based on the guidelines given in this chapter. Use Figure 3.1 as a guide.

3.4 Outline the primary results: Following on from Exercises 3.1–3.3, sketch the type of data you would present to the research sponsors on successful completion of your suggested research project.

3.5 Project timelines and budget: Using the research plan developed in Exercise 3.3, create a more detailed set of timelines showing maximum concurrent engineering to minimize the length of time required to complete the research. Could extra personnel or equipment decrease the time taken to conduct the research? Re-cast the project budget to reduce the time required to complete the project.

3.6 Using the results from Exercise 3.5, create a budget on a spreadsheet using the Table 3.1 template. Ensure that all aspects of the research have been included. Estimate the time required on various items of measurement equipment. Does your in-kind support value equate to the cash requirements of the budget?

3.7 Find a published journal paper in your discipline with more than four authors. From the paper list the positions and titles of all authors and from their biographies deduce what contributions were made to the paper by each of them. Note also the names and contributions of any persons acknowledged at the end of the paper. With this list of research participants, can you find any research details in the paper which are not covered by the research team you have listed?

3.8   Funding sources: Find a recent published research paper which acknowledges one or more funding sources. Check the web for details of the aims and objectives of the research funding organizations. Comment on how this paper reflects the overall stated strategy of the funding sources.

3.9   Locate a successful research project proposal (your instructor might have some) and identify the research question, the research team, the budget, etc and so comment on how well the structure of this proposal fits with the overall research grant writing guidelines given in this chapter.

3.10   Download a set of weather data from the internet covering the temperature and atmospheric pressure over a four day period. Present the data using 2D and 3D plots, and so deduce if the weather conditions are trending either higher or lower over this four day period. (Possible web sites include http://www.bom. gov.au/climate/data/ and http://www.silkeborg-vejret.dk/english/ regn.php).

3.11   Numerical modelling: Find a paper in which numerical modelling has been used to verify the experimental results. Comment on the differences between the experimental and modelling results. Have the authors commented on the accuracy of the experimental and modelling procedures? What suggestions do you have to improve the quality of the modelling reported in the paper?

References

Keywords: research grants, research strategy, engineering reports, grant writing, research funding, concurrent engineering, numerical techniques

[1] Blackburn, T.R., *Getting Science Grants: Effective Strategies for Funding Success*, San Francisco, CA: Wiley, 2003.

[2] Meador, R., *Guidelines for Preparing Proposals*, Boca Raton: CRC Press, 1991.

[3] Friedland, A.J. and Folt, C.L., *Writing Successful Science Proposals*, 2nd edition, New Haven, CT: Yale U, 2009.

[4] Eastman, C.M., *Design for X: Concurrent Engineering Imperatives*, New York, NY: Springer, 1996.

[5] Jonson, R., Freund, J. and Miller, I., *Probability and Statistics for Engineers*, 8th edition, Boston: Prentice Hall, 2011.

[6] Marder, M.P., *Research Methods for Science*, Cambridge: Cambridge University Press, 2011.

[7] Katz, M.J., *From Research to Manuscript: A Guide to Scientific Writing*, 2nd edition, New York, NY: Springer, 2009.

[8] National Association of Testing Authorities, Australia, http://www.nata.asn.au/, accessed December 2012.

[9] International Organization for Standardization, http://www.iso.org/iso/home.html, accessed December 2012.

[10] Sadiku, M.N.O., *Numerical Techniques in Electromagnetics*, Boca Raton: CRC Press, 2000.

[11] Bittnar, Z., *Numerical Methods in Structural Mechanics*, London: ASCE Press, 1996.

4

Statistical analysis

4.1 **Introduction**

The challenge in research is to prove observations and conclusions 'beyond reasonable doubt' (a common legal phrase). While most engineering research is numerically based and numbers are a prime outcome, some research is qualitatively based. In the latter research, one common method of analysis is to convert qualitative results to numbers and to use statistics to deduce the reliability of the outcomes and the conclusions. For this reason, regardless of the research methodology, researchers must pay particular attention to the use of statistics and the measures of uncertainty (experimental and modelling) in their research methods and in the development of conclusions.

It is expected that graduate engineers have already taken one or more courses in statistics. Readers are therefore referred to their undergraduate courses and the many reference books and internet references and software resources available (e.g. Matlab and MS Excel). This chapter provides a somewhat different approach to the standard undergraduate statistics textbooks [1–5], in order that engineers planning to undertake a research project have a fundamental understanding of statistics.

Engineers commonly over-specify requirements in order to ensure the reliability of their designs and production methods. Evaluations of the probability of failure, reliability calculations, and the inclusion of failsafe design are commonly used by practising engineers. For this reason, research which generates outcomes which do not anticipate failure (whether a catastrophic failure or simply a deviation from a specification) are unlikely to be broadly acceptable to the engineering industry and its clients. As a consequence, research outcomes which do not anticipate the probability of failure are better described as scientific investigations rather than engineering research.

Example 4.1 **Design and cost**

A bridge is designed to carry the weight of 40 vehicles. In constructing the bridge, the company might produce a design assuming 60 vehicles in addition to the one based on 40 vehicles, and perform a cost comparison for the two scenarios. If the cost differential is small, the structure might be better built for the higher vehicle loading. In doing so, the company is covering the possibility of less than perfect material properties, the impact of ageing on the load bearing capacity and the possibility of an extra heavy vehicle on the bridge.

As the primary objective of engineering is the improvement of the human condition, all engineering disciplines have an ethical obligation to ensure human safely. This is obviously evident in the biomedical engineering field where implantable devices and life-saving and life-maintaining devices might be the research outcomes. However, as all engineering projects support human existence, all branches of engineering must take this aspect of their research outcomes seriously.

This chapter investigates the concept of the appropriate level of accuracy required for research outcomes to be acceptable in human society. Some of the material might have been covered in secondary school and the early years of an undergraduate degree in engineering, but it remains of fundamental importance in the assessment of research outcomes. A failure to engage in the simplest of statistical and error analysis reflects very badly on the quality of the research, the research team and the conclusions. Thus statistics and error analysis must play a role in all engineering research projects.

4.2 Sources of error and uncertainty

Before discussing measurement errors which can't be controlled (randomly occurring errors), it is important for researchers to understand errors that can be controlled – usually through calibration – or at least predicted before the measurement strategy is implemented. These errors are described as *systematic errors* but can also occur due to the limitations of the measurement system. Most specifications of measurement systems relate to a linear response, however, the linearity of the response must also be quantified so that the error can be quantified over the measurement range of interest. Here are some of the common definitions used in sensors, transducers and measurement systems [6].

The *linearity* of a measurement system is defined in terms of its nonlinearity, that is, the difference between the true response when compared to the ideal linear response. As all measurement systems have a finite range of operation, it is necessary to define the *dynamic range* over which the system can be described as linear. This is illustrated in Figure 4.1, where the input to a measurement system is plotted on the *x*-axis and the output is plotted on the *y*-axis. The minimum value obtained from a measurement system is limited by such effects as the noise level in the instrumentation, and the maximum range can be limited by the maximum input and output voltages of the instrument. If the input is less than the range minimum, no output will be observed. It is possible that the instrument will be damaged (for example if the input has the wrong polarity, then the input amplifier can be

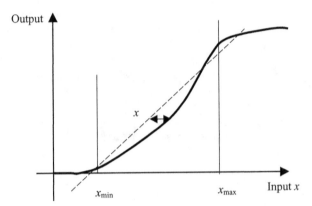

FIGURE 4.1 A linear approximation to the response of a measurement system is defined in terms of the ratio $\Delta x/x_{max}$ of the maximum difference Δx between the actual response (continuous line) and the ideal linear response (dashed line) and the maximum range x_{max}.

damaged). If the input range is exceeded then the output might be simply the maximum output value possible or, in the worst case, the measuring instrument might be destroyed.

The *resolution* of a measurement system is defined as the smallest change in input that can be detected at the output. In an analog system, the resolution can change if the system is nonlinear, but will remain constant when the system is linear. In a digital system the resolution is assumed to be linear when measurements are made within the range of the instrument.

Example 4.2 **Resolution**

A fixed range four digit digital volt meter has a maximum readout of 1.999 V. The resolution of this meter is therefore 1 mV. If the signal being measured changes by less than 1 mV, then no change will be observed in the display. The resolution of the voltage meter is 0.001 V.

The *sensitivity* of a measurement system is defined as the change in output divided by a small change in input. If the system has a linear response, then the sensitivity of the system is a constant over the working range. In Figure 4.1, the sensitivity is the slope of the straight line approximating the relationship between the output and input.

Example 4.3 Sensitivity

The change in the resistance of a wire $(\Delta R/R)$ under strain ε_l can be used to measure strain. A nichrome strain gauge is quoted as having a sensitivity S. This is the factor used to convert the relative change in resistance to units of strain. Thus the strain $\varepsilon_l = (\Delta R/R)/S$.

The *accuracy* of a measurement system is described in terms of the likely error. Errors can be the result of intrinsic problems with the instrumentation (calibration problems, nonlinear effects, imperfect components or couplings to the parameters being measured, etc) or external influences such as temperature, humidity, and materials which are present despite the best cleaning efforts and environmental control of the samples (often described as interferences).

Example 4.4 Interference effects

The intrinsic DC conductivity of ice is difficult to measure because the conduction across the surface is much greater than conduction through the crystalline bulk material.

The acoustic emissions from a rock core under uniaxial compression are often generated by the surface roughness of the ends of the cylindrical sample in contact with the jaws of the hydraulic frame rather than fracturing in the main body of the sample.

The compressive and shear strength of soils require the removal of samples and so the *in-situ* structure is compromised before testing. The compaction and water content of the soil can be changed significantly and the test outcomes cannot be applied to the *in-situ* measurements.

The *absolute error* is the difference between the measured value and the true value. Unfortunately there is no simple method of determining an absolute error. Repeated measurements most likely will produce the same absolute error. Improved calibration and the use of standard materials and components can reduce these systematic errors.

All measurement systems are subject to randomly occurring noise. For electronic systems this might be caused by the intrinsic properties of the electronics (a thermal noise level, contact noise effects, mutual coupling between adjacent conductors, etc) or external sources (for example coupled noise on the power and earth lines, the cables and the sensor itself). In Chapter 3 more than one sample was tested. It is possible that the samples under test might have slight differences in properties being measured,

Example 4.5 **Sample selection**

In a research project designed to measure the walking movements (gait) of adults, it might be necessary to exclude those who have suffered an injury in their foot or leg. Some might have slightly different leg lengths or balance problems which might influence the measurements. The research team must decide if these adults should be included in the measurements.

In a soil shear strength measurement at a building site, it will be necessary to test samples from a number of different locations around the site and to take samples at different depths. There might be a significant change in soil properties across the site. This will have implications for the building foundations.

In measuring the uniaxial compression strength of concrete samples it might be necessary to take samples at different times during the pour of a large slab. This is because the time of mixing can influence concrete strength and there might be variations in the volumes and size fractions of the components used in the mixture.

which are reflected in differences in the measured values. This is not a measurement error. Rather it is a different problem in which the method of selecting samples can play a major role in the measurements obtained. A clear scientific method of obtaining representative samples becomes important [7]. As the research outputs can be highly dependent on the sampling method, the research team must outline the method of sampling as part of the research plan.

4.3 One-dimensional statistics

A research paper reports a distance measurement of 10.5 m. The implication of this simple statement is that:

The measurement accuracy is 10.5 ± 0.05 m;

The measurement instrument has been calibrated;

The measurement instrument is capable of resolving measurements to this accuracy.

It also implies that a measurement of 10.7 m is significantly different from the result stated. Multiple measurements were introduced in Section 3.4. It is important to understand that the reporting of a single measurement contains inherent information about the accuracy of both the measurement and the measurement system.

One can extend this interpretation of a single number to include the statistical distribution. The histogram given in Figure 3.3 is an example where multiple measurements on the same system resulted in many slightly different results being recorded.

A common measure in statistics is the 5% probability estimate 'rule of thumb'. That is, 5% of all measured values will lie outside this range of values centred on the mean value. Conversely 95% of the measurements will lie within this range. This probability value is a measure of the random, symmetrical distribution of measured values about the mean value. Assuming a normal (random) distribution about the mean value μ, less than 5% of the measurements will lie outside the range of plus or minus two standard deviations (σ) away from the mean. On average, 2.5% of the population will have values greater than $\mu + 2\sigma$ (the upper tail of the population) and 2.5% of the population will have values smaller than $\mu - 2\sigma$ (the lower tail of the population). Clearly in situations where a 5% probability of error is unacceptably large, then a smaller probability might be mandated and the statistical analysis must be scaled accordingly.

Example 4.6 Acceptable probabilities and risk

An explosives company has a detonation device for large scale mining operations. A 5% probability of detonation is mandated by the mine operator. There is a greater than 50% chance that one or more charges will remain undetonated in a set of 100 explosive charges. This is likely to be unacceptable.

An artificial heart pump is used to replace the natural human heart. If there is a 5% chance that it will fail in the first two years of service, then this would be regarded as unacceptable by most people in the community.

The challenge of statistics is to define the levels of uncertainty based on measurement error and probabilities related to randomly distributed values. These are named *random errors* and are very different from *systematic errors* which result from some bias in the measurement technique (e.g. a calibration error).

Formally, the mean μ and standard deviation σ of the total population (sometimes referred to as the global population) includes those members of the population for which there is a measured value as well as those members of the population where the values are not measured. These parameters are defined by:

$$\mu = \frac{\sum\limits_{i=1}^{N} x_i}{N}, \tag{4.1}$$

$$\sigma = \sqrt{\frac{\sum\limits_{i=1}^{N} (x_i - \mu)^2}{N}}, \tag{4.2}$$

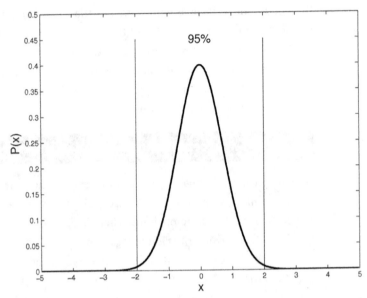

FIGURE 4.2 The normal distribution of an infinite population ($N \rightarrow \infty$) when the mean value $\mu = 0$ and the standard deviation $\sigma = 1$. The y axis indicates the probability P(x) of obtaining a value from a single measurement which lies within a particular range of x values. The probability of obtaining a value of x in the range $\mu - \infty < x < \mu + \infty$ is unity, and the probability of obtaining a value of x in the range $\mu \pm 2\sigma > x > \mu \pm 2\sigma$ is 0.95.

where x_i is the ith value of the population, N is the total number in the population and i is the counter between 1 and N. The mean value μ is a single value which represents the entire population. It is determined by minimizing the root mean square error (RMS) between the mean and all data points in the population. The mean value is NOT necessarily a measured value or a member of the population. The normal distribution is a continuous mathematical function which describes an infinitely large, unbiased, randomly distributed population (see Figure 4.2). In this figure the x-axis has been scaled so that the mean value is zero ($\mu = 0$) and the standard deviation is unity ($\sigma = 1$).

As it is rarely possible to make measurements on every member of the population and no population can be infinite in size, the only option is to approximate μ and σ using a smaller (hopefully representative) sample of the population. This is called the sample population. If there are n measurements, the sample population has n members and the mean and standard deviation of the sample population are given by equations similar to (4.1) and (4.2). The different symbols are used to indicate the difference. The mean of the sample population is $\overline{x}$ and the standard deviation is S given by:

$$\overline{x} = \frac{\displaystyle\sum_{i=1}^{n} x_i}{n}, \tag{4.3}$$

$$S = \sqrt{\frac{\displaystyle\sum_{i=1}^{n} (x_i - \overline{x})^2}{n - 1}}. \tag{4.4}$$

The approximate relationships between μ and σ and x and S are:

$$\mu \cong \chi, \tag{4.5}$$

$$\sigma \cong S \sqrt{\frac{n}{n - 1}}\, S. \tag{4.6}$$

From the normal distribution, the best estimate of the mean value of the population is x and there is a 95% chance that any value x measured lies inside the range $\mu - 2\sigma < x < \mu + 2\sigma$. Note that while μ can have any value (i.e. either positive or negative), the standard deviation is always positive (i.e. $\sigma > 0$).

The normal distribution shown in Figure 4.2 has been scaled such that $\mu = 0$ and $\sigma = 1$. Usually the mean value of a population is not zero and the standard deviation is not unity. In order to calculate the probability of obtaining particular values in a normally distributed data set, it is common to transform that data set so that μ and σ are zero and unity respectively. Any value of *x* can be scaled using the transformation equation:

$$z = \frac{x - \mu}{\sigma}.$$ (4.7)

The *z* value (called the *z*-score) representation of the *x* value now has a normal distribution with $\mu = 0$ and $\sigma = 1$. The probability that the measured value of *x* lies within the normal distribution of the data set can be calculated from this distribution. The conversion from a *z*-score to a probability and the reverse (i.e. from the *z*-score to *x*) can be undertaken using the statistics functions in programs such as Matlab (see the function zscore and normcdf) and Excel (see the function normsinv). Alternatively the probability values can be read from a printed set of *z*-score tables found in most undergraduate statistics textbooks.

Before accepting the value returned by these functions for the probability it is important that the meaning of the returned value(s) is/are well understood.

As the normal distribution is a continuous function, the probability of obtaining a single, precise value is infinitely small. The normal distribution will only return a finite value of the probability *P* between zero and unity if a range of *z* is specified (i.e. an upper and lower limit of *z*). If a range of values is not specified, then the function might assume the range is from zero to the value, from the value to infinity or from minus infinity to the value, etc.

As the normal distribution is symmetrical about the mean, 50% of the values lie above the mean and 50% lie below. As a result a *one-tailed* probability estimate is half the probability of a *two-tailed* probability estimate. A one-tailed probability estimate is the probability that the value lies between minus infinity and the specified *z* score or between the *z* score and plus infinity. The two-tailed probability estimate is the probability that the absolute value lies between minus infinity and the negative *z* score, and between the positive *z* score and plus infinity. Thus the value for the two-tailed probability is twice the value obtained for the one-tailed probability value.

Example 4.7 z-score probability

What is the probability that a value of $x = 7.76$ is part of a previously measured data set where the mean value was determined to be 6.2 and the standard deviation was determined to be 2.1?

Commentary: You must calculate the z score of the new x value using (4.7). Thus $z = 0.74$ and so the x value lies within one standard deviation of the mean. It is highly likely that this new value is a member of the original data set.

There are many situations where the normal distribution cannot be used for probability calculations. The most likely are:

▨ The mean value is very close to zero and negative values are not possible in the data set.
▨ The distribution is skewed about the mean. This is defined numerically as the skewness of the population.
▨ The $\mu \pm 2\sigma$ range of values do not contribute 95% of the probability. This is numerically defined as the kurtosis of the population.

The appropriateness of using the normal distribution to describe a population can be tested mathematically. The *skewness* of the distribution is calculated from the data set using the equation:

$$skewness = \frac{\sum_{i=1}^{N} (x_i - \mu)^3}{(N-1)\sigma^3}. \tag{4.8}$$

The skewness for a distribution similar to a normal distribution should be approximately zero. A negative value of skewness indicates that a greater percentage of the population lies in the lower values. A positive value of skewness indicates that a larger number of values lies in the higher values. For example, if there is a lower limit to the data and the mean value is relatively small, then it is likely that the skewness is positive. A significant value of skewness might also be explained by the influence of an additional variable which contributes to the population set.

The *kurtosis k* is calculated using:

$$k = \frac{\sum_{i=1}^{N} (x_i - \mu)^4}{(N-1)\sigma^4}. \tag{4.9}$$

For a normally distributed data set (i.e. one which is similar to a normal distribution) $k \cong 3$. If k is significantly greater than 3, this means that the distribution has a narrower peak when compared to a normal distribution.

Example 4.8 **Moments in statistics**

Note that Equations (4.4), (4.8) and (4.9) have an increasing power (2, 3 and 4 respectively). These parameters are sometimes referred to as the second, third and fourth moment in the data distribution.

If a data set does not follow the parameters of a normal distribution, what must be done and what conclusions can be drawn? Firstly, there are some additional investigations and analyses that might be performed on the data set.

Could it be that the parameter measured is normally distributed using another scale (for example a log scale, a power series, etc). This can be investigated by re-evaluating the mean, standard deviation, skewness and kurtosis using a different function such as a logarithmic relationship (e.g. $\log x_i$) or a power law (e.g. x_i^n) rather than x_i. Should this technique be successful, it can give some fundamental insight into the nature of the measurements being made. Further advanced techniques are available but these are beyond the scope of this book.

Alternatively some other transformation techniques can be applied so that a standard 5% error assessment can be used in follow up calculations. In a skewed data set, the upper 2.5% limit will be different from the lower 2.5% limit. The techniques outlined in the following sections can still be applied but must be carried through with different upper and lower error bounds.

In MS Excel, the skewness and kurtosis can be determined using the descriptive statistics tool. In Matlab the functions are kurtosis and skewness.

4.3.1 Combining errors and uncertainties

Once several parameters have been determined experimentally and their associated errors determined using the 5% probability concept, some additional mathematical processing might be required in which the different parameters and their associated errors are combined to calculate the final value of interest and the associated error. There are some simple rules for combining errors which are based on the least squared error analysis used to calculate the mean value with Equation (4.3).

If two values are to be added or subtracted: $x_i \pm 2\sigma_i$ and $x_j \pm 2\sigma_j$, then the result y is given by

$$y = (x_i \pm x_j) \pm 2\left(\sigma_i^2 + \sigma_j^2\right)^{1/2}. \qquad (4.10)$$

Note that all of the units must be the same. That is the units of y must be the same as the units of x_i and x_j.

If two values are divided or multiplied: $x_i \pm 2\sigma_i$ and $x_j \pm 2\sigma_j$, then the result y is given by

$$y = \frac{x_i}{x_j} \pm \frac{x_i}{x_j} 2\sqrt{\left(\frac{\sigma_i}{x_i}\right)^2 + \left(\frac{\sigma_j}{x_j}\right)^2}. \qquad (4.11)$$

Here the units of y and x_i, x_j do not have to be identical. These two mathematical expressions are based on the RMS analysis and so are statistically rigorous and should be used in combining data with their associated errors.

If the calculation involves addition, subtraction, multiplication or division, then Equations (4.10) and (4.11) must be applied in the same sequence as the calculations. If the mathematical operation is nonlinear, then an input number with equal limits will result in an output number with unequal limits.

Example 4.9 **Nonlinear error calculations result in unequal errors**

You are required to determine $A = \sin \theta$ where $\theta = (50 \pm 5)$ degrees. The result is $A = 0.766 + 0.053$ & $- 0.059$.

You are required to determine $B = \log d$ where $d = (1.5 \pm 0.3)$. The result is $B = 0.176 + 0.079$ & $- 0.097$.

An inspection of (4.10) shows that if one error is much larger than the other, then the smaller error can be ignored. In (4.11) if the relative error of one parameter is much larger than the other, then the smaller relative error can be ignored. These approximations can simplify data analysis.

4.3.2 **t test**

In some experimental investigations it is important to know if two populations are likely to be sample populations selected from the same global population. This type of one-dimensional question can be addressed using a *t test* (sometimes referred to as *Student's t test*). The test is most applicable when the standard deviations are very large in comparison to the likely changes or differences between the two mean values.

There are two main types of t test:

- *The paired t test* is one in which the same population is tested twice to determine if there has been a change in the overall population. The reference to students is because the test evaluates the probability that the class of students has demonstrated a statistical improvement in marks relevant to all students in a class. More generally it is a method of determining if there is a statistically significant change in the population after an intervention. A simple mean and standard deviation calculation will not show a significant change if the change is likely to be significantly smaller than the standard deviation measure of the population.

Example 4.10 **Paired t test**

A group of 20 aluminium poles of different sizes is weighed immediately following manufacture. The same poles are weighed after six month's exposure to the environment. The paired t test will give the probability that there is a significant change in the weight of the poles.

A group of 25 employees is weighed. A fitness trainer is asked to work with the group to lose weight. Six months later the same group of people is weighed again. The paired t test will provide a probability estimate that a significant change in weight has taken place. The value of the fitness training can be established.

A simple mean and standard deviation calculation will not show a significant change if the change in weight is likely to be significantly smaller than the range of weights in the population.

▨ *The unpaired t test* is one in which two different populations are measured to determine if there is a difference between the two populations. In this case the two populations are unrelated and the number of samples can be different in the two sample sets.

Example 4.11 **Unpaired t test**

The concrete strength tests in Ghana high rise buildings need to meet an international specification. A set of measurements from a sample in England was used for comparison. What is the probability that these two sets of measurements are identical? The question can be resolved using a t test.

Populations of 25 ten-year-old girls in Sweden and 30 ten-year-old girls in Denmark are weighed. The objective is to see if there is a difference in weight between the two populations. The probability that there is a difference between the two populations can be established using an unpaired t test.

The t test can be conveniently evaluated using the MS Excel function ttest-paired and ttest-unpaired. In Matlab the functions are ttest and ttest2 for the paired and unpaired data sets respectively.

4.3.3 ANOVA statistics

In Section 4.3.1 one-dimensional statistical methods were outlined. In Section 4.3.2 two different populations were compared using the t test. If there are more than two dependent data sets, then the previous techniques are inadequate under many circumstances. For example, if many repeat measurements are made of a number of members of the population, the *ANOVA* statistical methods allow the calculation of probability estimates for three or more data sets. As with the t test, this method of analysis can determine statistically significant differences when the standard deviations in the parameters are much larger than the difference between the populations.

Example 4.12 **ANOVA probability testing**

Populations of 25 ten-year-old girls in Sweden and 30 ten-year-old girls in Denmark are weighed and their height is also measured every year over a 20 year period. The objective is to see if there is a relationship between height and weight between the two populations over time. The probability that there is a difference between the two populations tracked over the years can be established using a two-factor ANOVA test.

The ANOVA test can be simply evaluated using the MS Excel function ANOVA – two-factor with replication and ANOVA – two-factor without replication. In Matlab the functions are anova1 and anova2.

4.3.4 Example: 1D analysis of flood height measurements

Because of the vast size of Australia, the sparse population in regional and remote areas, the large numbers of river crossings and the absence of a reliable source of electric power, painted river level markers are common (see Figure 4.3). The local landholder will be asked to read the river levels during a rain event which might cause flooding. While there are some electronic river level gauges connected to the telephone network on major rivers, they require mains power to operate. In storm conditions in remote areas the power is unreliable. Electronic river height gauges are expensive to install, they are subject to vandalism and can be washed away during flood events. For these reasons the painted flood gauges are common.

The objective of making river level measurements is to determine the rise during a flood event. A reliable estimate of the rise in river height will allow an accurate prediction of how much flooding will occur downstream in a residential area. Before the rain which caused the flood, the river level gauge (see Figure 4.3) was read at $h_1 = 0.70 + 0.05$ m. The error was assessed based on the poor resolution of the marker. At the river level peak, a series of measurements h_2 was made at the same gauge, but the viewing distance was much larger because the additional water causes an increase in width of the river. In addition, the river is flowing more rapidly and the varying wave height makes the visual measurements imprecise and error prone.

A series of 13 height measurements ($n = 13$) was made over a relatively short time in an attempt to obtain an accurate value of the river height. Table 4.1 shows the measurements and the times when the readings were made.

FIGURE 4.3 A painted wooden post serves as a river height gauge in Queensland, Australia.

The initial assumption is that this is a one-dimensional measurement – that is, the height is independent of the time over which the measurements were taken. The mean value and standard deviation of the data can be calculated using Equations (4.3) and (4.4). Using the river height gauge shown in Figure 4.3, it is quite difficult to estimate the river height to within 1 cm. Reading the gauge by eye from a distance will inevitably lead to some random errors.

If the variations are due to random noise, then the data should be normally distributed. The mean value and two standard deviations from the mean value are $h_2 = 1.62 \pm 0.04$ m. The skewness (Equation (4.8)) is -0.23 and the kurtosis (Equation (4.9)) is 2.42. The skewness and kurtosis are approximately 0 and 3

TABLE 4.1 River height measurements h_2 at the flood peak.

Time (hrs)	h_2 (m)
8:55	1.58
8:57	1.60
8:58	1.63
9:01	1.63
9:05	1.62
9:06	1.62
9:07	1.62
9:08	1.63
9:11	1.63
9:12	1.66
9:13	1.62
9:15	1.62
9:18	1.61

respectively and so it is reasonable to assume that the data are normally distributed.

The same data set can be reviewed to see if there is a significant rise in the river level based on these statistics. The z-score for h_1 given the h_2 statistics is $z = (0.7 - 1.62)/0.018 = -51$. The probability that h_1 is a member of the population of h_2 is extremely small. We can therefore assume that the river has risen significantly. The reader might say this result is obvious after considering the mean and standard deviations. This has now been verified statistically.

The increase in river height can be calculated using Equation (4.10). The result is

$$h_2 - h_1 = (1.62 - 0.70) \pm (0.04^2 + 0.05^2)^{1/2}$$
$$= 0.92 \pm 0.06 \text{ m.}$$

From this analysis the following conclusions can be drawn:

- The river height has risen by an average of 0.92 ± 0.06 m.
- There is a 2.5% chance that the river will rise to a height greater than 0.98 m. This is based on a one-tailed probability calculation using the calculated error as two standard deviations from a normally distributed mean.

What has not yet been discussed is whether the time is an important parameter in these measurements. There is a possibility that the river height is a function of time. A two-dimensional analysis is required.

4.4 Two-dimensional statistics

Chapter 3 contained a brief introduction to the possibility that a particular one-dimensional measurement might be dependent on the time when the measurement was taken. This implies that the mean value of the population might continue to change as time passes. If the population without this time dependence was normally distributed, then what is the effect on the distribution if the mean value changes with time? This requires a further analysis based on a two-dimensional approach to the problem. It is not appropriate to announce a result of one value and its 5% error bounds. In statistics the same methods apply regardless of whether the data are one-, two- or multi-dimensional. The level of analysis depends on the research team's view of the experimental situation. The error might be initially assumed to be normally distributed around the mean, but in two-dimensional data, the mean might continue to change. For illustration it will be assumed that there is a possibility that the mean value of the measurements changes with time.

The method of statistical analysis is called correlation (least squares regression analysis or correlation analysis). Is the measured parameter correlated with time? An initial linear fit can be used to determine if there is a relationship between the measured parameter and time. That is, the river height h might be a linear function of time t. This can be written as

$$h = mt + h_0, \tag{4.12}$$

where h_0 is the river height at $t = 0$ and m is the rate of change of the height with time (the slope of the line).

The correlation coefficient r (sometimes referred to as the Pearson correlation coefficient) is a statistical parameter used to evaluate the quality of the linear relationship:

$$r = \frac{n \sum_{i=1}^{n} t_i h_i - \left(\sum_{i=1}^{n} t_i \right) \left(\sum_{i=1}^{n} h_i \right)}{\sqrt{n \left(\sum_{i=1}^{n} h_i^2 \right) - \left(\sum_{i=1}^{n} h_i \right)^2} \sqrt{n \left(\sum_{i=1}^{n} t_i^2 \right) - \left(\sum_{i=1}^{n} t_i \right)^2}}$$

$$= \sqrt{1 - \frac{Q_R}{Q_m}}, \tag{4.13}$$

where t_i are the time values corresponding to the measurements x_i, Q_R is the sum of the differences squared between the measured values h_i and the fitted straight line defined by Equation (4.12). This can be written as:

$$Q_R = \sum_{i=1}^{n} (h_i - (mt_i + h_0))^2, \tag{4.14}$$

and

$$Q_m = \sum_{i=1}^{n} \left(h_i - \overline{h} \right)^2 \tag{4.15}$$

is the sum of the difference squared between the measured values h_i and the mean of the measured values $\overline{h}$.

There are many computer programs which calculate r and r^2. In MS Excel the function is correl and in Matlab it is corrcoef.

- When $r = 1$, the correlation is perfect and no random error is present in the data. The parameter x increases linearly with time.
- When $r = -1$, the correlation is perfect and no random error in the data is present. The parameter x decreases linearly with time.
- When $r \cong 1$, the data are said to be highly correlated or strongly correlated with time.
- When $r \cong 0$, the data are said to be uncorrelated and there is no relationship between the measurements and the time at which they are taken.

▦ When $r = 0.975$, then $r^2 = 0.95$. This means that 95% of the relationship with time is linear and 5% is random variations in the data.

Returning to the 5% probability interpretation, it is possible to say that if $r > 0.975$ or $r < -0.975$, then there is a strong linear relationship between the measurements x_i and the time t_i at which they were made. While all of the measured data points have been used in this statistical analysis, there might be a logical reason to disregard one or more data points. This option is discussed later in this chapter.

The methods of approach used to acquire the data might include various sampling strategies. For example, let us consider which is the better strategy:

Option 1: Measure a large number of data points in a short period of time (say 24 points in 1 hour), and then repeat the measurements every day for 20 days.

Option 2: Make one measurement every hour for 20 days.

Option 3: Measure 240 points on the first day and then 240 data points on day 20.

All three measurement strategies give the same number of data points. Is there a statistically preferable option?

Option 1: Calculate the sample mean and standard deviation using Equations (4.3) and (4.4), check that the data from one day are normally distributed, and plot the mean and standard deviation on a scatter plot with error bars as a function of time. A linear fit to the mean values together with the correlation coefficient calculation will reveal the strength of the relationship between the measured parameter and the time of measurement. Assuming a normal distribution, it is also possible to use a z-score calculation to check the probability that the mean at the end of the sampling period is significantly different from the mean of the first day data set. This might offer conclusive proof that a change in the mean value has occurred.

Option 2: This strategy would be to scatter plot the data versus time and perform a linear regression analysis between these two parameters. The value of the regression coefficient *r* will indicate the strength of the dependence. The coherency of the points on the scatter plot will give a visual clue about the strength of the relationship, but a rigorous statistical analysis requires that the probability of a change over time is quantified and assigned a probability value.

Option 3: This strategy would be to perform a normal distribution analysis and *z*-score comparison between the two data sets. The increased number of points in each data set will give a more statistically reliable result for the mean and standard deviation between the two groups of data.

Regardless of which data acquisition option is used, the calculation of a linear correlation coefficient using all data points is a useful first step in the statistical analysis. This analysis does not require a regular time interval between the consecutive data measurements. If a scatter plot of the data shows a consistent monotonic change with time, then a nonlinear correlation analysis should be performed.

A nonlinear data analysis uses RMS error minimization to fit the data to a function suggested by the researcher. One method is to 'linearize' the data using a strategy similar to that described in Chapter 3. The alternative method is to fit a nonlinear function directly to the data. Excel provides options for nonlinear fits to two-dimensional data sets (e.g. power series, log, exponential and trigonometric functions are presented as options). Matlab uses the function polyfit to fit higher order polynomials to a data set. Note that option 3 will not provide robust statistical support for nonlinear function fitting as there are essentially only two times when data are available. Option 2 will provide the strongest data set for nonlinear fitting as all parameter values are treated equally, regardless of the time of the measurement.

In many experimental cases, there is no simple mathematical relationship between the measured parameters. In this case the research team might develop an 'empirical' mathematical relationship. That is a relationship which can accurately describe the data and can be used to interpolate within the measured data set, and even extend the interpretation beyond the measured data set, even though the relationship might not fit a theoretical concept. In the time series of measurements discussed in this section, it is possible to estimate the value of x at times much greater than the value recorded at the last measurement time x_N. This is referred to as extrapolation. The quality of the functional fit can still be assessed using RMS error minimization, and the normalized RMS error. This process can be described as mathematical optimization and this is the subject of Chapter 5.

4.4.1 Example: 2D analysis of flood height measurements

Returning to the flood height measurements reported in Table 4.1, the possibility that the data are time dependent must be assessed. If there is a general trend that the river is still rising or is falling, the one-dimensional analysis in Section 4.3.4 will have failed to yield an accurate view of the situation. This might mean that the river height peak has either passed or has yet to arrive and the total river height peak is larger than that predicted from the one-dimensional analysis. How can this theory be investigated in a statistically rigorous manner? The following procedure can be used:

Step 1. Generate a scatter plot of the data and fit a linear trend line through it (see Figure 4.4). There appears to be a slight increase in river height over the measurement time. However, it is clear from a visual inspection that the data are not strongly linear.

The figure suggests that the river height increased at the start of the measurement period and then decreased towards the end of the period. The linear trend line has a slight positive slope (the slope is 0.065 m/hr) indicating that there is a possibility that the

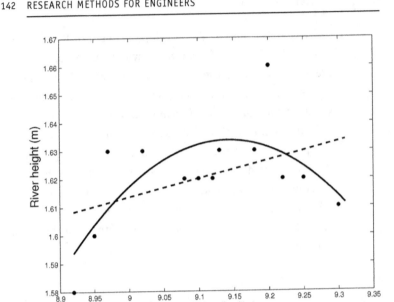

FIGURE 4.4 River height plotted as a function of the time of day during a flood (Table 4.1 data). The dashed line indicates the best fit straight line (linear regression line), and the continuous line is the quadratic best fit.

river level might be continuing to increase, even after the measurements had stopped. The correlation coefficient is very low ($r^2 = 0.175$) and so the statistical significance of the linear trend is weak. It might be logical to conclude that the onset of the river rise is much faster than the decrease in level after the flood peak has passed. A further statistical evaluation is required to assess the importance of the large data value of 1.66 m at time 9:12 am. Is this value of statistical significance? If so, then the river peak might be much larger than the one-dimensional analysis has predicted.

Step 2. Perform a nonlinear least squared error fit to the data. The rise and fall across the time measurements perhaps can be fitted to a quadratic equation, and the correlation coefficient might be larger indicating an improved fit of the function to the data. The line is included in Figure 4.4. The quadratic polynomial fit results in $r^2 = 0.495$. This value is an increase over the linear fit, but the functional fit is still not particularly strong.

Following the suggestion that the river rise and fall might have different time scales, a cubic polynomial might show a stronger relationship. With $r^2 = 0.498$ the cubic polynomial fit has marginally improved the RMS error.

Increasing the order of the polynomial fit should increase the value of r^2 because more unknowns are involved. In the limit, a polynomial of order n, where n is the number of data points, should result in a perfect fit as all data points will lie on the line. Unfortunately this adds little to an explanation of the situation.

Step 3: This requires a further examination of the highest data point of 1.66 m recorded at time 9:12 am. In this case one can evaluate the probability that this point is a member of the measured data set. If this can be proved statistically, then this single value may be excluded from the data set, and the mean and standard deviation can be recalculated. The z-score for this excluded point can be calculated to determine the probability that it is a member of the data set.

The new mean value and its associated error without the 9:12 am data point result is $h_2 = 1.62 + 0.03$ m. While the change from the previous mean does not appear significant, if the value is correct, then the impact on the analysis and the community downstream can be very severe. The z-score for $h_2 = 1.66$ m is $z = -2.87$ and the probability that this is part of the random variations in the population is 0.008 or 0.8%. This result suggests that the probability of this single data point being part of the total population is less than 1%. Perhaps this much higher river height reading can be attributed to human error or an abnormally high wave passing the gauge.

The major conclusions from this analysis can be:

- The maximum value of the river height has increased by approximately 0.92 m.
- The river height measurements are time dependent.
- The rise in the river level is quicker than the fall.
- The measured data set includes one suspect measurement which should be removed from the calculations.

■ There is no statistically strong mathematical function which can be used to describe the rise and fall of the river level.

The flood management and warning authority must make a decision about the flood height at this point in the river. The data will be used in the hydrographical model to estimate the possibility and severity of flooding downstream from this measurement point.

4.5 Multi-dimensional statistics

As two-dimensional analysis is an extension of one-dimensional statistics, so also multi-dimensional analysis uses an approach similar to that described in Sections 4.3 and 4.4. The difference is that one or more additional parameters must be included in the RMS error minimization.

For example, if the measured parameter y is influenced by two different parameters x_1 and x_2, then a linear relationship can be written as a modified version of Equation (4.12). That is:

$$y = a_0 + a_1 x_1 + a_2 x_2. \qquad (4.16)$$

One can calculate the values for a_0, a_1 and a_2 by inverting the following set of 3×3 equations (called the normal equations):

$$\sum_{i=1}^{n} y_i = a_2 \sum_{i=1}^{n} x_{2i} + a_1 \sum_{i=1}^{n} x_{1i} + na_0,$$

$$\sum_{i=1}^{n} x_{1i} y_i = a_2 \sum_{i=1}^{n} x_{2i} x_{1i} + a_1 \sum_{i=1}^{n} x_{1i}^2 + a_0 \sum_{i=1}^{n} x_{1i},$$

$$\sum_{i=1}^{n} x_{2i} y_i = a_2 \sum_{i=1}^{n} x_{2i}^2 + a_1 \sum_{i=1}^{n} x_{1i} x_{2i} + a_0 \sum_{i=1}^{n} x_{2i}. \qquad (4.17)$$

The subscript i refers to the recorded values of y, x_1 and x_2. The multiple linear regression coefficient r_{xy} is still given by the equation

$$r_{xy} = \sqrt{1 - \frac{Q_R}{Q_M}}, \tag{4.18}$$

where Q_R is given by

$$Q_R = \sum_{i=1}^{n} [y_i - (a_0 + a_1 x_{1i} + a_2 x_{2i})]^2, \tag{4.19}$$

and Q_M is similar to Equation (4.15),

$$Q_m = \sum_{i=1}^{n} (y_i - \overline{y})^2. \tag{4.20}$$

Equations (4.17) through to (4.19) can be easily modified to include any number of variables and any functional form.

If there are three variables then Equation (4.16) must include an additional unknown constant and the normal equations (4.17) must include an additional summation line. Thus there will be four equations to solve for four unknowns. The calculation and interpretation of the correlation coefficient calculated using Equation (4.18) will remain the same.

The generalised form of (4.19) can be written as:

$$Q_R = \sum_{i=1}^{n} [y_i - f_i)]^2, \tag{4.21}$$

where f_i is the fitted function calculated at the measurement points i. A rigorous multi-dimensional statistical analysis can therefore be conducted using a set of normal equations and Equations (4.18), (4.20) and (4.21) can be used to calculate the correlation coefficient.

4.5.1 Partial correlation coefficients

Consider an experiment in which three parameters are measured α, V and T. If the two-dimensional correlation coefficients between the three parameters ($r_{\alpha V}$, $r_{\alpha T}$ and r_{VT}) are calculated, it is possible to minimize the influence of one variable in the correlation coefficient on the other two parameters using the partial correlation coefficient $r_{\alpha T(V)}$ defined by

$$r_{\alpha T(V)} = \frac{r_{\alpha T} - r_{\alpha V} r_{VT}}{\sqrt{1 - r_{\alpha V}^2}\sqrt{1 - r_{VT}^2}}. \qquad (4.22)$$

Equation (4.22) can be used to evaluate the correlation coefficient between α and T without the effect of V. In an experiment where there is no control over the dependent variables, Equation (4.22) can be used to remove the influence of one of the parameters while evaluating the correlation coefficient between the other two parameters.

4.5.2 Example: Diggability index

The effectiveness of blasting rock at a quarry for road base is measured by a 'diggability index' (D). It is measured as the amount of rock which can be fed to a crusher over one hour, i.e. the rock crusher throughput. It is suspected that D is influenced by rain within the 5 days preceding the blast. This index is also known to be influenced by the number of explosive charges used in the blast. The quarry engineer wishes to test the significance of this hypothesis given the data in Table 4.2.

The first task is to convert the non-numerical data into numerical data. This can be done in many different ways. For example, a simple binary system could be used where Y = 1, and N = 0. If recent rain is likely to have a more significant effect on D, then the rain index (a measure of rain significance) could be computed by calculating the parameter Y/(day number).

Table 4.3 is a modified version of Table 4.2 with a numerical rain index R obtained by using the equation:

$$R = \sum_{i=1}^{5} \frac{R_i}{d_i}, \qquad (4.23)$$

where $R_i = 1$ for Y and $R_i = 0$ for N, and d_i is the number of days before the blast.

TABLE 4.2 A data record for quarry blasting. Y/N indicates whether rain fell in the days (1–5) before the blast. D is the diggability index which was measured by rock crusher throughput.

Charges	D	Days before the blast				
		5	4	3	2	1
35	0.4	N	N	N	N	N
86	0.75	N	Y	Y	N	N
75	0.6	Y	N	N	Y	N
98	0.8	N	N	N	Y	Y
20	0.35	N	N	Y	Y	Y
25	0.28	Y	N	Y	N	N
80	0.59	N	Y	Y	Y	N

TABLE 4.3 Numerical scores for the rain index R for the data in Table 4.2.

c	D	R	d_i				
			5	4	3	2	1
35	0.4	0	0	0	0	0	0
86	0.75	0.58	0	1	1	0	0
75	0.6	0.7	1	0	0	1	0
98	0.8	1.5	0	0	0	1	1
20	0.35	1.83	0	0	1	1	1
25	0.28	0.33	0	0	1	0	0
80	0.59	1.08	0	1	1	1	0

It is now possible to compute the linear correlation coefficients between c, D and R. The three results are:

$$r_{Dc} = +0.97,$$
$$r_{RD} = +0.23,$$
$$r_{Rc} = +0.16.$$

As predicted, r_{Dc} is close to 1 indicating that there is a strong linear correlation between D and the number of charges used in the blast. The correlation between D and R can be calculated using the partial correlation coefficient $r_{DR(c)}$ using Equation (4.22). The result is

$$r_{RD(c)} = +0.31.$$

The conclusion is that the rain index R does not appear to have a significant influence on the diggability index D. The correlation coefficient is positive implying that D is higher when R is higher, but the small value of $r_{RD(c)}$ suggests that there might be other factors influencing D. Further research might be appropriate and additional factors might be included in the statistical analysis.

4.6 Null hypothesis testing

The objective of research projects is to provide an answer to a carefully phrased research question. Various types of research questions are discussed in Chapter 3, and commonly the question raised will use one of the following words: How? When? Why? What? These words are then followed by a hypothesis.

For example the question relating to the statistical analysis given in Section 4.3.4 might be:

'What is the maximum river height during this flood?'

The hypothesis could then read:

'The river height increased by less than one metre during the period 8:55 to 9:18 am.'

The null hypothesis could be written:

'The river height did not reach its maximum value during the period 8:55 to 9:18 am.'

Note that the null hypothesis is not an exact complement of the hypothesis. The alternatives to the null hypothesis could relate to the time value or the change in river height value, but these two possible null hypothesis statements are restrictive and do not clearly address the central question.

The statistical analysis presented in Section 4.4.1 uses the quadratic functional fit to the field data to indicate that the maximum river height was 1.63 m at 9:10 am and, following that time. the river height started to decrease.

The research question for Section 4.5.2 might read:

'What is the effect of rain on the rock crushing throughput in quarry blasting?'

The hypothesis would be:

'If rain falls on the quarry site before a blast, the blast effectiveness is increased.'

The null hypothesis could be written:

'Rain has no effect on blast effectiveness.'

The statistical analysis presented in Section 4.5.2 revealed $r_{RD(c)} = +0.31$ which means that 10% ($= 0.31^2$) of the variation in diggability can be attributed to the effect of rain. The remaining 90% of the variation in the diggability cannot be attributed to the effect of rain. Arguably the null hypothesis has been supported statistically and so there is little support for the claim that rain has an influence on the effectiveness of rock blasting.

4.7 Chapter summary

Engineering researchers must engage in the statistical analysis of their results in order to prove their conclusions are valid. This can be clearly stated by calculating the probability that each of the conclusions presented in a final report or scientific paper is valid.

The research outputs can range from the evaluation of one number (with appropriate error analysis and statistics), an analysis of a single data set (mean, standard deviation, skewness and kurtosis), a comparison between one population before and after an intervention (paired t test), a comparison between two unrelated populations (unpaired t test) and multi-variable analysis (ANOVA) in which many individual objects are measured many times over a period of time. The t tests and ANOVA tests are particularly important if the standard deviation about the mean value is very large in one or more sets of the data.

There are a number of computer based tools (e.g. MS Excel and Matlab) which allow a very quick statistical analysis to be performed on data sets [3, 8]. A failure to use these statistical methods indicates a lack of professionalism in conducting the research and may lead to the results and conclusions being challenged.

The analysis presented here is a brief summary of a very large number of statistical techniques available. Novice researchers might need to consult the statistical textbooks for a deeper understanding of the statistical methods outlined in this chapter and to investigate other techniques which might be relevant to their particular research problem.

When developing a research proposal, the statistical methods of analysis should be proposed so that the sponsoring body has some confidence that the research outcomes are realistic and the conclusions are strong. This initial planning contained in the research proposal should impact on the type and volume of the data gathered.

Exercises

4.1 Sensor statistics: Review the specification sheets of three experimental instruments used to make measurements related to your engineering discipline. Briefly summarize the following user requirements:

Dynamic range;

Sensitivity;

Linearity;

Calibration requirements;

Calibration procedure.

4.2 Normal equations: Assume you conduct an experiment measuring y_i, t_i, x_i and s_i. Your data should fit the following equation:

$$y = a_0 + a_1 t + K x^3 + C \log s.$$

Write out the normal equations you would use to solve for the constants a_0, a_1, K and C.

Hints: You will need four equations to calculate these four unknowns.

Convert your raw data to a linear scale (i.e. use x^3 and $\log s$ data values in the equations).

Use Equation 4.17 as a template.

4.3 Statistical reporting: Rework the statistics from an old undergraduate laboratory assignment or other data set. Write out the research question, hypothesis and null hypothesis. Using a more detailed statistical analysis write out a conclusion from the experiment. You should seek a number of different experiments and perform a 2D and multi-dimensional analysis on the data. If appropriate, calculate the partial correlation coefficient. What are your conclusions?

4.4 Statistical review: In your engineering discipline review a published paper which includes a statistical analysis. Write a brief report on the statistical methods used. Can you suggest an improved statistical analysis? Suggest some additional parameters that might have been measured during the data acquisition stage and so explain how you would analyse the total data set to deduce the influence (and statistical significance) of these additional measurements.

4.5 Statistical review: In your engineering discipline review a published paper which does not include a statistical analysis. Write a brief report suggesting how the data published in the paper can be supported through an improved statistical analysis. Can you calculate a probability that the result in the paper is correct?

References

Keywords: measurement theory, measurement error, statistical analysis, normal distribution, z-score, probability, correlation coefficient, least squares error analysis, sampling methods, Matlab statistics, Excel statistics, t test, ANOVA

[1] NIST/SEMATECH, *e-Handbook of Statistical Methods*, 2003, http://www.itl.nist.gov/div898/handbook/, accessed January 2013.
[2] Veneziano, D., *Probability and Statistics in Engineering*, lecture notes, MIT OpenCourseWare, 2005. http://ocw.mit.edu/courses/civil-and-environmental-engineering/1-151-probability-and-statistics-in-engineering-spring-2005/lecture-notes/, accessed January 2013.
[3] Triola, M.F., *Essentials of Statistics*, 4th edition, Boston, USA: Addison Wesley, 2011.
[4] Marder, M.P., *Research Methods for Science*, Cambridge, UK: Cambridge University Press, 2011.

[5] Johnson, R., Freund, J. and Miller, I., *Probability and Statistics for Engineers*, 8th edition, Boston, USA: Pearson, 2011.

[6] van Putten, A.F.P., *Electronic Measurement Systems*, New York: Prentice Hall, 1988.

[7] Stuart, A., *Basic Ideas of Scientific Sampling*, 2nd edition, London: Charles Griffin & Co Ltd, 1976.

[8] Salkind, N.J., *Excel Statistics: a Quick Guide*, Los Angeles, CA: SAGE, 2011.

5

Optimization techniques

5.1 Introduction

Almost all engineering design problems are multi-parameter problems. For example a solid object has a three-dimensional shape in addition to the material properties relevant to the application. Material properties might include bulk mechanical properties, electromagnetic properties and chemical properties in addition to surface treatments such as a passivation layer in microelectronics and environmental protection coatings for pipelines. The simplest structure, a three-dimensional rectangular shaped piece of material, will require engineering design decisions about the length, width and thickness of the object. If it is hollow (i.e. it contains an air cavity), then the wall thickness and the method of joining different sections (e.g. welding, adhesive bonding, soldering, screws, etc) must be decided. In some cases the external

Example 5.1 **Standard measures**

In the building industry the relative proportions of the length, width and thickness of a house brick are similar world-wide.

The size of aluminum soft drink containers is the same world-wide.

Drill bits, spanners (wrenches) and screwdrivers commonly come in standard sizes.

The USB (universal serial bus) connector is mandated by international standards.

dimensions of an object are determined by engineering standards or previous commonly accepted sizes.

While the physical dimensions of a house brick might be determined by an international agreement (even if this is unofficial), the strength of a wall made from these bricks is dependent on both the size and the mechanical properties of the brick. Bricks can be made from mud, fired clay and concrete and may be solid or hollow, and can contain indentations and holes. An analysis of the strength of the wall can be undertaken by computer modelling or theoretical analysis. However, the construction of a wall is not limited to the bricks alone. The mortar (cement) used between the bricks will add to the effective size of the brick, and the strength of the mortar will contribute to the overall strength of the wall, particularly in directions perpendicular to the plane of the wall. An effective engineering design should take into account the total cost of the wall, the time required for construction and the environmental cost in mining and processing the raw materials and transport to the site. The environmental cost can be determined using a full life cycle analysis (LCA) [1], and there are commercial software suites and databases available to undertake LCA assessments in every discipline of engineering.

Example 5.2 **Environmental impacts and LCA**

Is it better environmentally to construct walls using prefabricated concrete slabs or bricks? Both construction techniques require some manufacturing off-site and then transport to the building site. The erection of the wall will be quicker using the slab technique but will require more mechanical lifting to place the walls in position. A brick wall is likely to be thicker than a slab wall and so the volume of material will be less using slabs. All of these considerations must be taken into account in the life cycle analysis.

The design engineer must find the best design from this large array of independent variables. An engineer engaged in research might seek to find a design which satisfies all of many different requirements relating to:

- Physical size;
- Construction/fabrication materials;
- Monetary cost and material availability;
- Environmental cost and material transport.

Mathematically the cost of the wall can be represented as a 'cost function' C which includes all of the relevant components and the strength of the wall. The best design (the optimized design) will have the minimum cost function providing that a minimum set of structural requirements is met. Optimization is the mathematical process of defining all of the parameters which will lead to the best possible design taking into account all of the engineering, the monetary cost and the environmental costs. Such calculations provide excellent opportunities for engineering research for both novice and experienced engineering research teams.

One method of approach to design the best possible wall is for the members of the engineering research team to use their professional experience to suggest which materials will produce an optimal design from a structural and environmental perspective. The possible designs are then tested using experimentation, computational modelling and theoretical analysis. The cost function is calculated and compared with the cost function derived from the alternative designs. *Forward modelling* is the process of calculating the cost for a set of known parameters. This method of approach can be quite slow as the research team attempts to fine tune the design by changing the parameters to obtain an optimal result. The result that they decide is optimal might not be the best result, and there is no method of determining if the result is optimal.

The alternative is to use a mathematical method which solves the problem many times in a logical, strategic manner to determine the best possible design. For each new design the cost function is recalculated until the best possible result has been determined. This process is referred to as *inverse modelling* or *optimization.*

The cost function does not have to be dimensionally sound from a mathematical point of view – that is, as the parameters can have different units (metres, dollars, volts, kPa, tonnes, dB, seconds, etc), the cost function can be formed from the sum of all contributing factors and each contributing factor can be weighted in order of importance to the final result. There will be range limits on all of the parameters as the optimization routine must not be allowed to give answers where the parameters lie outside these ranges. For example the physical size of an object should not exceed the maximum handling capacity of the user and the maximum fabrication length possible in the factory.

Example 5.3 Wider impacts on a cost function

The design of a 400 km pipeline might ideally use a single pipe of length 400 km. However, the manufacture, transport and positioning of a 400 km length of pipe will be regarded as impossible. The design team needs to consider a balance between the length of the individual sections and the number of joins along the length of the pipeline.

The design of an undersea optical fibre communications cable will have similar challenges to the pipeline. Unlike the pipeline, an optical cable can be coiled so that larger lengths can be transported easily.

The outcome from an optimization routine can be surprising as the design is not necessarily guided by physical theory or intuition. The result can be unusual shapes and configurations. This is one of the exciting research aspects of using mathematical optimization.

The calculation of the linear correlation coefficient defined by Equations (4.17)–(4.20) is based on minimization of the RMS error between the experimental data and a theoretical mathematical function. The optimal values of the n constants in the function given by Equation (4.16) are determined by the inversion of an

$n \times n$ matrix (Equation (4.14)). The list of requirements outlined in this section assumes that there is no mathematical function which can be applied to the problem and so the matrix inversion method of approach outlined in Chapter 4 is not possible.

In this chapter the challenges of optimization will be demonstrated using two fitted parameters contributing to the cost function. This allows the various optimization techniques explored to be demonstrated visually (i.e. on a two-dimensional plane). The same techniques can be expanded to multiple dimensions using similar equations, however, it is impossible to clearly represent the path to the optimized solution in n-dimensional space.

5.2 Two-parameter optimization methods

The methods commonly used to optimize a design problem are most simply illustrated if only two parameters are involved. This allows the optimization process to be represented on three-dimensional plots where the third dimension is the value of the cost function. These techniques are illustrated using a curve fitting task related to sports engineering.

In many sports activities, a push or swing is a common characteristic. For example a tennis racquet swing, a cricket bat swing, a baseball bat swing, running, swimming, boxing and ballet all have movements which start at rest and finish at rest [2, 3]. Similar equations apply to other engineering disciplines such as the movement of pistons and pumps in mechanical engineering and electrons in alternating current circuits and electromagnetic waves. In sporting movements, the position of a limb such as an arm or leg, or implement such as a bat, racquet, stick, as a function of time $D(t)$ can be modelled using a sinusoidal squared equation over the range $t = 0$ to $t = 1 = \pi/2\omega$:

$$D(t) = L \sin^2(\omega t), \tag{5.1}$$

where L is the amplitude of the movement (angular displacement or linear distance), ω is the angular frequency of the movement and t is the time. Confining this example to linear

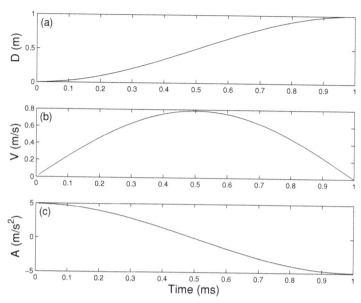

FIGURE 5.1 Variation in (a) distance D, (b) velocity V and (c) acceleration A as a function of time t for a linear displacement following Equations (5.1)–(5.3).

movement, then D is the distance and $0 < D < L = 1$ for $0 < t < \pi/2\omega$. This function is illustrated in Figure 5.1a.

Equation (5.1) satisfies the requirements that the distance travelled changes monotonically from 0 to L, and the velocity is zero at the start and the end of the movement. Differentiating Equation (5.1) gives the velocity $V(t)$ where

$$V(t) = \frac{dD}{dt} = 2L\omega\cos(\omega t)\sin(\omega t) = L\omega\sin(2\omega t). \qquad (5.2)$$

At $t = 0$ and $t = \pi/2\omega$, $V(t) = 0$. This is shown in Figure 5.1b. The acceleration $A(t)$ is given by

$$A(t) = \frac{d^2D}{dt^2} = 2L\omega^2\cos(2\omega t). \qquad (5.3)$$

The function in Equation (5.3) is illustrated in Figure 5.1c.

The acceleration can be measured using accelerometers attached to the implement or limb of the player [2–3]. From these acceleration measurements, a fitted curve will allow the determination

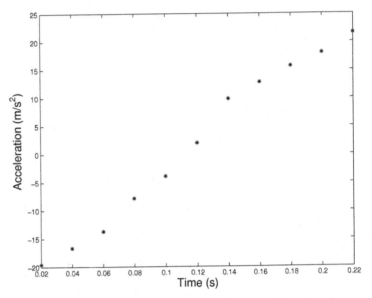

FIGURE 5.2 Experimentally determined acceleration $A(t_i)$ values from a linear movement. The acceleration changes from negative to positive values. This indicates that the positive axis of the accelerometer is in the opposite direction to the movement. The sample rate is 50 per second.

of the constants L and ω and so a complete interpretation of $D(t)$ and $V(t)$.

The objective is to find the best (optimal) values for D and ω which are determined by a comparison of the function given in Equation (5.3) and the measured data from the accelerometer. The numerical value to be minimized is called the cost function. In this case the cost function $C(D, \omega)$ is the RMS value given by

$$C(D, \omega) = \sum_{i=1}^{N} (A_i - A(t_i))^2, \qquad (5.4)$$

where A_i are the experimental measured values at time t_i, $A(t_i)$ are the values determined using Equation (5.3) and N is the total number of sample points. The equation has the same form as Equations (4.14) and (4.21).

For this example, assume that the acceleration is sampled at 50 samples per second (i.e. at 0.02 s intervals). The experimental data are illustrated in Figure 5.2. As the start time is not known the data will only be fitted to the two parameters if an offset time t_0 is included in Equation (5.3). The new equation becomes

$$A(t) = 2L\omega^2 \cos 2\omega(t + t_0). \tag{5.5}$$

The acceleration A at times $t = 0$ and $t = \pi/2\omega$ is not zero. This is physically not possible as all time derivatives of $D(t)$ must be zero at $t = 0$ and $t = \pi/2\omega$. There must be a lead time before the acceleration reaches its maximum values at the start and the end of the movement. For simplicity, the monotonic portion of the acceleration will be used in this example.

In two-dimensional optimization t_0 will be fixed and L and ω will be varied to minimize C. The optimization will be improved if t_0 is also added to the list of parameters but for illustration purposes only 2D optimization will be implemented in this example. Inspection of the experimental data indicates that $t_0 = 0$ is a reasonable estimate. Two-parameter optimization methods are readily adapted to three-parameter optimization, and this is discussed later in this chapter.

The cost function C must be calculated multiple times (Equation 5.4) using the estimated values of L and ω, until a minimum C value is found. Optimization techniques can be either *guided* or *unguided* optimization. The unguided techniques rely on a predefined strategy to determine the next set of parameters L and ω. A value of C is then calculated and compared to the previous cost function values. The algorithm terminates when all predefined values have been tested. The values of L and ω which contribute to the minimum value of C are the optimized solution to the problem.

Guided optimization techniques use recently derived values of C to select (i.e. to guide) the next set of parameters L and ω, and C is again determined. The next values of L and ω are determined from the previous values of C. The algorithm terminates when no further improvement in C can be achieved. This coincides with the optimal values of L and ω and the minimum value of C.

There are many optimization strategies [4, 5]. Some are based on biologically inspired strategies (e.g. the genetic algorithm [6, 7], ant colony [8], particle swarm [9]), and some are inspired by physical processes (such as simulated annealing [10]). In the following subsections, four techniques are illustrated. Sequential uniform sampling is an exhaustive search method in which all possible solutions are tested [11]. The Monte Carlo method is a purely random search method. The simplex method and gradient methods are based on vector calculus concepts.

The quality of the solution depends on a number of preset limiting conditions and the (as yet unknown) shape of the cost function:

- The range of the parameters L and ω (called the solution space);
- The step size of L and ω (the resolution);
- The number of minimum values of C in the solution space.

It is possible that there is more than one minimum value of C in the solution space. If so, there is more than one solution to the optimization routine. It is essential to locate the best possible solution which coincides with the minimum value of C. If there is no minimum value in the solution space, then the range of parameters must be increased or one must conclude that there is no optimal solution. When using a guided optimization routine, it is possible that the first solution obtained will coincide with a minimum value of C which is not the overall minimum value of C in the solution space. This is called a *local minimum*. For these reasons, all guided optimization routines should be run more than once using randomly generated start values.

5.2.1 Sequential uniform sampling

The most obvious approach to optimization is to test all possible values of L and ω. It is still necessary to choose the range limits for L and ω, a relatively small increment for both L and ω, and sequentially step through all possible values of both parameters in the solution space range. This technique is referred to as sequential uniform sampling [11] (Figure 5.3). In order to explore and demonstrate the efficacy of the various algorithms, a full view of

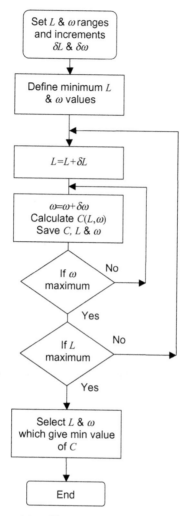

FIGURE 5.3 Sequential uniform sampling algorithm for two parameter optimization. C is the cost function calculated from the RMS error between the fitted equation and the experimental data. The two parameters to be optimized are L and ω. The routine checks C for all possible values exhaustively. The optimum values of L and ω correspond to the minimum value of the cost function.

the cost function over the parameter range of interest is presented. This is an outcome of the process of sequential uniform sampling.

The flow chart in Figure 5.3 outlines the algorithm used for sequential uniform sampling. Commonly the technique will be applied more than once. The first attempt should use a coarse grid with large incremental steps in each of the parameters. Once a possible optimized parameter range has been obtained, a smaller range with smaller incremental steps can be used to improve the resolution of the solution. There are flaws in this method of approach in that 'local minimum' values in the cost function might not be defined and investigated. It is possible that the best minimum cost function value will be missed if the initial incremental values are set too large for all parameters.

Figure 5.4 shows the cost function C results for all possible values of L in the range $0.003 < L < 0.3$ with a step size $\delta L = 0.003$ (100 points) and ω in the range $0.3 < \omega < 10$ with a step size $\delta \omega = 0.3$ (30 points). The offset time was set to $t_0 = 0$. The total number of calculations required for an exhaustive search is 3000. For problems involving a complex, time-consuming numerical calculation, the total time to run the method will be very large indeed.

Example 5.4 **Complex computational models**

We wish to calculate the impact of rapid deceleration of the brain resulting from a sports related impact. A model of the human head might require more than 10 M voxels if the anatomical details are included precisely.

We wish to calculate the radar reflection of a ship. The surface of a three-dimensional model of a ship will require more than 1 M pixels.

A finite element solution of these structures will take a significant period of time. An optimized solution may take many hours or even days for an exhaustive optimization routine to be completed.

Figures 5.4a and 5.4b show that the variation in C around the optimal solution (the minimum value of C located at $\omega = 7.0$ rad/s and $L = 0.207$ m) is quite small for step size changes used for L and ω.

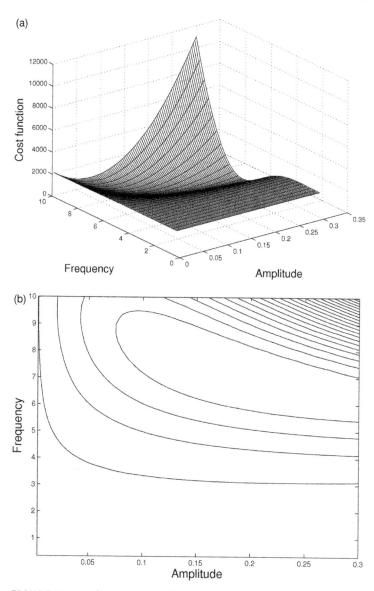

FIGURE 5.4 C values for all possible L and ω values in the range $0.003 < L < 0.3$ and $0.3 < \omega < 10$: (a) the mesh plot; (b) the contour plot. The objective is to find the values of L and ω which coincide with the minimum value of C.

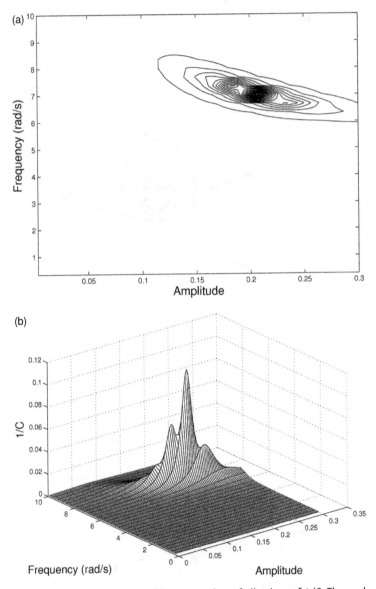

FIGURE 5.5 (a) Mesh plot, (b) contour plot, of all values of $1/C$. The peak corresponds to the minimum value of C.

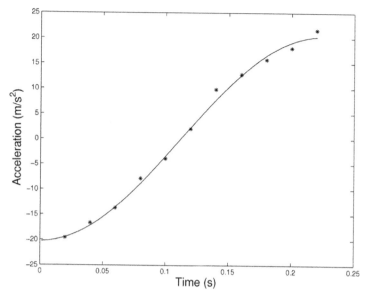

FIGURE 5.6 Fitted acceleration data $A_i(t)$ using Equation (5.3). The values of L and ω were determined from the minimum value of the cost function C.

An alternative presentation method is to consider the inverse values of C (see Figure 5.5a and 5.5b). This is referred to as the *fitness function* and must be maximized to obtain the optimized solution. This converts the shallow minimum C value to several well-defined peaks. The cost function $C(L, \omega)$ therefore has several well-defined solutions but only one dominant solution. An effective optimization routine must ensure that the solution converges to the best possible and is not trapped in one of the local minima. The values of L and ω can be determined from the highest peak in $1/C$. The optimized solution ($L = 0.207$ m and $\omega = 7$ rad/s) is plotted together with the experimental data in Figure 5.6. A good match is evident through visual inspection.

Figure 5.7 shows the variation in C with every iteration following the sequence in Figure 5.3. It is clear that at some point the optimized values of L and ω have been passed and yet the program continues to calculate every possible set of values. For this reason, this method of optimization is described as unguided.

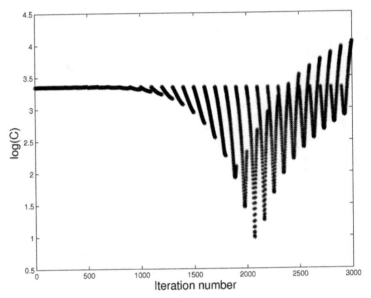

FIGURE 5.7 The variation in the cost function C plotted for every value of L and ω. The iteration number is plotted along the x axis and follows the sequence of calls to the function.

For improved resolution of the fitted parameters L and ω, the offset parameter t_0 can be optimized. Figure 5.8 shows the sequential variation in C using the optimized values of L and ω and sequentially changing t_0. An improved value of C is obtained.

While this looks like the best possible solution, the individual optimization of parameters will not provide the best solution. This is because of the interdependence of the three parameters. In this case the optimized solution must have an improved solution as all three parameters L, ω and t_0 are solved simultaneously rather than independently. There is a significant relationship between ω and t_0. For this reason, there is no certainty that the optimal values of L, ω and t_0 have been derived accurately using this method. With $t_0 = 0.008$ ms, a new cost function minimum has been obtained and so the optimum values of L and ω should be recalculated.

The number of calculations required to optimize all three parameters using sequential uniform sampling becomes 300 000 if 100 values of t_0 are tested. In this case the forward solution

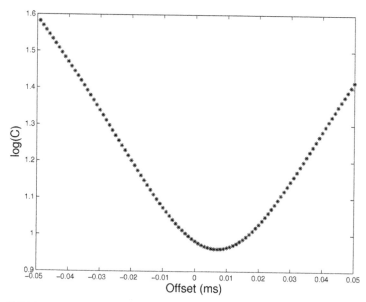

FIGURE 5.8 Variation in C as t_0 is varied systematically. The optimum values of L and ω derived using $t_0 = 0$ s were used in these calculations.

is not particularly difficult or time consuming; however, if the forward solver is slow (for example 15 minutes per solution), the total run time for the optimization procedure will be very long (approximately 77 hours). It is clear that this optimization method will have improved accuracy when the sample intervals for each parameter are small and the range is large. This is the reason why alternative optimization methods are sought for multidimensional problems.

5.2.2 Monte Carlo optimization

The Monte Carlo optimization technique is based on random selection of parameter values within the defined range [11]. Figure 5.9 illustrates the algorithm. For this example, the probability of randomly selecting values that are close to the correct value depends on the 3000 options available (this is based on the range and interval used in the sequential uniform sampling algorithm). The number of random samples (L_i and ω_i) tested with the cost function depends on the sensitivity and range of the parameters.

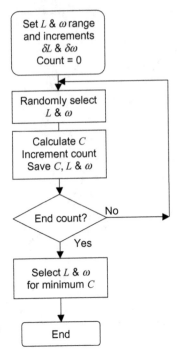

FIGURE 5.9 Monte Carlo sampling algorithm for two-parameter optimization. C is the cost function calculated from the RMS error between the fitted equation and the experimental data. The two parameters to be optimized are L and ω. The routine checks C for the randomly chosen values of each parameter until the predefined total number of samples has been completed. The optimum values correspond to the minimum value of the cost function.

An inspection of the 3000 points in Figure 5.7 reveals that less than 30 points (1% of the total data set) have $\log(C) < 2$. Using a Monte Carlo data set of 2% of the population (i.e. 60 from a total of 3000) the probability of obtaining $\log(C) < 2$. is very low.

If the search range for L and ω is restricted to ranges closer to the correct value, then the probability of gaining useful results increases significantly. Of course, this is only possible if an approximate solution is known. In this case the Monte Carlo method is used to improve the accuracy of the solution. The Monte Carlo method is an unguided optimization technique.

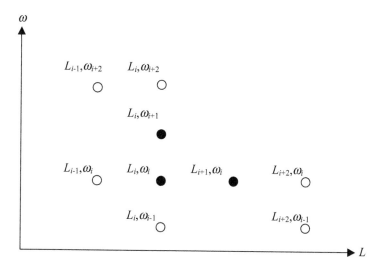

FIGURE 5.10 A simplex triangle defined in the two-dimensional solution space by the points (L_i, ω_j), (L_{i+1}, ω_j) and (L_i, ω_{j+1}) is shown by the black dots. Depending on the cost function values at each of these three points, the point with the maximum cost function is replaced by one of the open-circle positions.

5.2.3 Simplex optimization method

This optimization method employs a directed strategy to determine the minimum value of the cost function. The start point for L and ω in the range is selected randomly. The cost function for two adjacent points is calculated. The first is a $\delta\omega$ step in the ω direction from the start point. The second is a δL step in the L direction from the start point. These three adjacent points are called a *simplex* [11, 12], see Figure 5.10. In three-dimensional analysis the simplex has four points and in an N-dimensional problem the simplex has $N+1$ points.

The path towards the optimal solution is guided by comparing the cost functions for each of the three adjacent points in the simplex. From this information, a new point is selected by stepping in the direction of the minimum C value (away from the position of maximum C in the simplex). This new point is added to the simplex and the point with the highest C value is eliminated. This

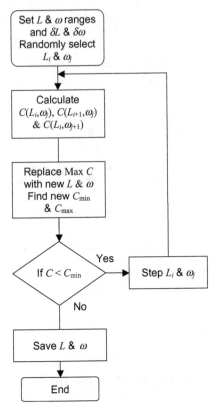

FIGURE 5.11 The simplex algorithm flow chart. The start point L_i, ω_j is selected randomly. The next two points in the simplex are determined. The cost function is calculated for all three points and the point with the maximum cost function is replaced. The process continues until a minimum value of the cost function C is determined.

forms a new three point simplex. The process is repeated until no reduction in C is observed. Figure 5.11 illustrates the simplex optimization algorithm.

The algorithm can be written as follows:

- If $C(L_i, \omega_j) > C(L_{i+1}, \omega_j) > C(L_i, \omega_{j+1})$ then $L_i, \omega_j \rightarrow L_i, \omega_{j+2}$;
- If $C(L_i, \omega_j) > C(L_i, \omega_{j+1}) > C(L_{i+1}, \omega_j)$ then $L_i, \omega_j \rightarrow L_{i+2}, \omega_j$;
- If $C(L_{i+1}, \omega_j) > C(L_i, \omega_j) > C(L_i, \omega_{j+1})$ then $L_{i+1}, \omega_j \rightarrow L_{i-1}, \omega_{j+2}$;
- If $C(L_{i+1}, \omega_j) > C(L_i, \omega_{j+1}) > C(L_i, \omega_j)$ then $L_{i+1}, \omega_j \rightarrow L_{i-1}, \omega_j$;

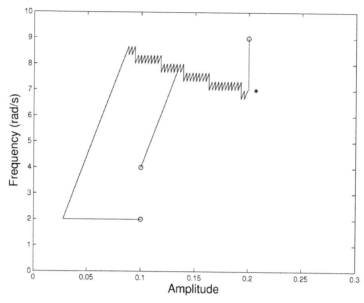

FIGURE 5.12 Using four different starting locations, the simplex algorithm tracks through the solution space until a minimum value of the cost function C is located. Note that the path can only take one of six possible directions.

- If $C(L_i, \omega_{j+1}) > C(L_i, \omega_j) > C(L_{i+1}, \omega_j)$ then $L_i, \omega_{j+1} \to L_{i+1}, \omega_{j-1}$;
- If $C(L_i, \omega_{j+1}) > C(L_{i+1}, \omega_j) > C(L_i, \omega_j)$ then $L_i, \omega_{j+1} \to L_i, \omega_{j-1}$.

The cost function is calculated for the new position on the L, ω axes. The process is repeated until no reduction in the cost function is possible.

The flow chart is given in Figure 5.11 and four typical paths are illustrated in Figure 5.12. Note that the algorithm defined in Figure 5.11 allows for six possible new points only. There are only three possible directions of the path with positive and negative directional options. Movement in another direction is achieved in a zigzag path and the progress towards the optimal solution is slowed. If the path is trapped in a local minimum, then the process terminates with a sub-optimal solution.

The start position L_i, ω_j is selected randomly. As there is a possibility that the path will be trapped in a local minimum value,

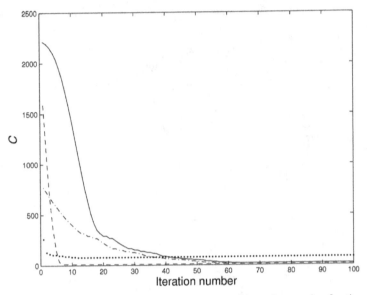

FIGURE 5.13 The variation in C as a function of iteration number for the four paths shown in Figure 5.12. The speed of convergence to the minimum value depends on the start position.

the complete algorithm must be run several times with different start positions in order to increase the probability that the global minimum cost function is obtained. This corresponds to the best possible values of the parameters and the best fit to the experimental data. The speed of convergence also depends on the start position. The variation in cost function for the four paths shown in Figure 5.12 is shown in Figure 5.13 as a function of the iteration number.

There are many many variations in this basic method of approach aimed at improving the speed of convergence.

5.2.4 Gradient optimization method

This method is similar to the simplex method, but uses a slightly more direct route to the solution by placing more emphasis on the relative values of the three points of the cost function in the simplex. The slope of the path through the solution space is calculated using the gradient vector $\underline{\nabla C}$ [11, 13], defined by:

$$\underline{\nabla C} = \frac{\partial C}{\partial L}\underline{x} + \frac{\partial C}{\partial \omega}\underline{y}, \tag{5.6}$$

where $\underline{x}$ is the unit vector parallel to the L axis and $\underline{y}$ is the unit vector parallel to the ω axis. This straight line path is the line of steepest descent. The slope of the line m is given by:

$$m = \frac{\partial C/\partial \omega}{\partial C/\partial L}. \tag{5.7}$$

The next position in the solution space is given by the new value of L and the equation:

$$\omega = m(L - L_{max}) + \omega_{max}. \tag{5.8}$$

The start point for the initial simplex is randomly selected. The maximum C value at position L_{max}, ω_{max}, for the simplex is defined at the start point. The value of C at the start point is then calculated using Equation (5.8). The path follows a straight line defined by the gradient given by $\underline{\nabla C}$ in Equation (5.6). When C ceases to decrease along this path, the simplex is again defined at that minimum point and a new gradient vector is calculated. The process continues until no further reduction in C is obtained. The best solution for L and ω identified at this point forms the best solution. It is possible that the path can become trapped in a local cost function minimum.

A flow chart for the gradient optimization technique is given in Figure 5.14. Figure 5.15 shows the positional change in the solution space and Figure 5.16 shows the convergence of the algorithm. Figure 5.15 shows that the start point of $L = 0.1$ and $\omega = 2$ results in a very large deviation in the first iterations, after which it rapidly converges on the solution. The start points at $\omega = 8$ and $\omega = 9$ follow the same path to the solution. The path which starts at $\omega = 5.5$ becomes trapped in a local minimum value.

In three of the four cases, the speed of convergence shown in Figure 5.16 is faster than the simplex method.

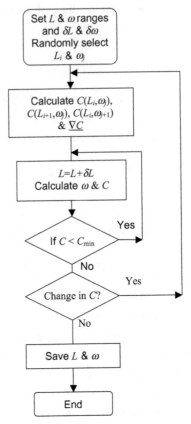

FIGURE 5.14 The gradient optimization algorithm. Using a randomly selected start position, the cost function for all points on the simplex is calculated. The gradient from these points defines the direction of the path through the solution space. When the cost function no longer decreases, a new simplex is defined and the process is repeated until no further decrease in C is possible.

As with the simplex method outlined in Section 5.2.3, not all solutions result in locating the position of the minimum value of the cost function in the solution space. It is therefore essential to run the algorithm a number of times using randomly located start positions. This can reduce the likelihood of the algorithm terminating in a local minimum. The optimal solution is that solution where the cost function is a minimum.

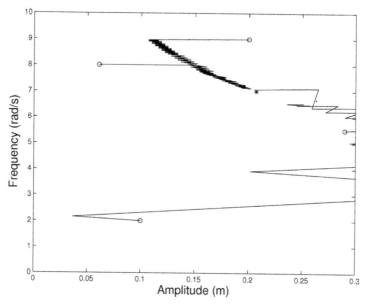

FIGURE 5.15 Using four different starting locations (○), the gradient algorithm tracks through the solution space until a minimum of the cost function C is located. The path takes a direction defined by the gradient calculated at the simplex. The path for start position $L = 0.1$ deviates off the scale but then recovers to locate the optimum solution (∗) quite rapidly.

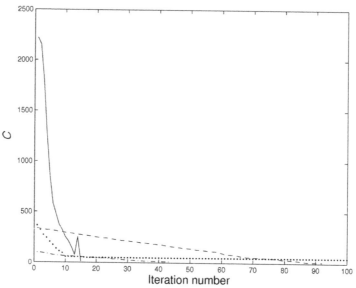

FIGURE 5.16 The variation in C as a function of iteration number for the four paths shown in Figure 5.12. The speed of convergence to the minimum value depends on the start position. Note that the dotted line corresponds to the start point of $L = 0.29$ which terminates in a local minimum.

5.2.5 **Summary**

The speed of convergence is significantly different for each optimization method. While sequential uniform sampling required 3000 calculations of C, the Monte Carlo method had a much smaller number of calculations but had a very small probability of achieving the optimal solution. The simplex and gradient methods converged to the solution more rapidly, but the number of calculations of C was highly dependent on the random start point. The quickest solution was achieved by the gradient method in 15 calculations.

With the optimal values for L and ω defined by this optimization process, the function D is uniquely defined using Equation (5.1). From this, the maximum velocity can be obtained. By comparison, a numerical integration of the acceleration data yields a velocity V with a constant, and a second integration to find the distance D results in an additional constant and a linear offset. The mathematical approach given by Equations (5.1) and (5.3) is far more efficient.

5.3 Multi-parameter optimization methods

The four methods outlined in Section 5.2 have demonstrated the algorithmic details for a two-dimensional investigation. The routines can all be modified simply to allow for an additional set of parameters to be optimized simultaneously. For example the offset parameter t_0 should be included to obtain a more precise fit to the theoretical relationship given by Equation (5.1). All optimization algorithms can also be used for such multi-parameter investigations.

Many of the commercial forward modelling computing codes have an optimization routine as part of the package. A common concern about these optimization routines is the time required for the solution of one model. If the time to complete one solution is large, then multi-parameter optimization routines can be extremely slow as the forward modelling routine must be implemented many times before the optimal solution is reached.

There is significant research interest in improving the speed of convergence of optimization routines in addition to the computational efficiency of the forward modelling codes.

Some optimization codes are freely available on the internet. Matlab has an optimization tool box and MS Excel optimization routines run with their solver box.

There are several additional strategies used to reduce the computational time required for multi-parameter investigations. If the software is licensed to run on multiple machines, then massively parallel computing techniques can be used to reduce the overall computational time. Another approach is to use a perturbation technique for the solution space. In this case, small changes to the parameters to be optimized can still be calculated, but the majority of the solution space is either unmodified or modified slightly. This can reduce the computational time required as the solution of the complete forward model is not required.

One common strategy is to use a very coarse optimization grid with a relatively small number of data points to reduce the solution space range where the cost function minimum is likely to be found. Once located, the routine can be run with restricted boundaries using a finer grid. The inherent danger here is that if the resolution is too coarse in the initial stages of the investigations, then the global minimum value might be missed and a sub-optimal solution will be obtained.

One of the more interesting possibilities with multi-parameter optimization is the opportunity to solve problems with more unknowns than experimental measurement points. This is possible

Example 5.5 **Curve fitting**

The electromagnetic surface impedance is a function of the depth and conductivity of horizontally stratified layers. In the simplest case of a single layer above an infinitely deep earth half-space, there are three unknowns: the conductivity of the upper layer, the conductivity of the lower layer and the depth of the lower layer. The complex surface impedance measured at a single frequency only provides two measurements – the real and imaginary parts of the impedance. Given the depth of penetration of the radiation and the limits on the conductivity it is possible to deduce the three parameters from the complex surface impedance using optimization.

if there is a well-defined mathematical theory, and there are well-defined limits and other constraints on the parameters being fitted. The optimization methods described in this chapter can be applied to such ill-defined problems. This approach is different from the curve fitting approaches used in Chapter 3 where the number of data points is commonly much larger than the number of unknowns.

5.4 The cost function

The selection of the cost function is of critical importance to all optimization algorithms. In the examples given in this chapter, the cost function was a single number – the RMS error between a set of experimental data and a theoretical function.

Consider the case of a design engineer seeking to minimize the weight W and maximize the strength S of a structural beam. The cost function might be defined as:

$$C = w_1 W + w_2/S, \tag{5.9}$$

where w_1 and w_2 are the weighting factors used to balance the importance of the two parameters. The weights can be modified to change the emphasis on one parameter relative to the other. It also allows the cost function to have no units as w_1 will have the units of inverse W and w_2 will have the units of S.

With most structural beam designs, an increase in strength is accompanied by an increase in weight. With this type of dependency, it is possible to define a set of optimal solutions, rather than one unique solution. Figure 5.17 is an example of a data set in which the best possible solution is defined as a curve across the best possible solutions for various W and S values. From Figure 5.17, a designer can choose a particular value for W and determine the best possible value for S. Given the nature of the curve, it is clear that increasing the weight W does not result in a proportional increase in the strength S of the beam. This curve is referred to as the pareto front or attainment line. If three

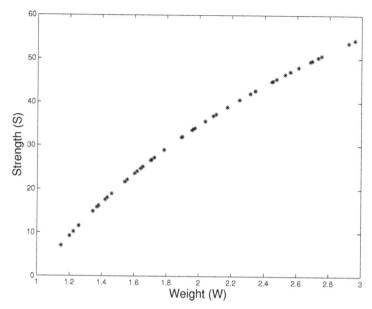

FIGURE 5.17 A hypothetical data set of the optimized solutions of a structure in which both the weight W and strength S have been determined. The optimized solutions form a line. The design engineer can choose which optimal solution is preferred.

parameters are used in the cost function the pareto front becomes an attainment surface. The concept is easily extended to multi-dimensional problems and the same optimization approach can be used.

Example 5.6 Cost function definition

The design of radio frequency identification tags (RFID) for mass production of more than 10M per production run is aimed at optimizing the design for minimum cost D, adequate material properties M, minimum size S, lower frequency f, and minimal environmental impact E assuming disposal in a landfill site. The cost function C might be written as:

$$C = w_1 D + w_2 M + w_3 S + w_4 E,$$

where w_1, w_2, w_3, and w_4 are the weighting factors.

The cost function can include a wide variety of parameters including those which are discrete (such as an option to use one of several different materials), non-scientific parameters such as the dollar cost and environmental impact parameters which arise from complete product life cycle assessments.

5.5 Chapter summary

Multiparameter optimization is a mathematical technique used to find the best possible solution to a complex design problem. The objective is to minimize the cost function which includes all of the parameters relevant to the best possible design. The speed of convergence can be increased through the use of guided algorithms such as the simplex and gradient methods.

The cost function can be designed to include many different objectives and the outcome can be a multidimensional optimization 'surface' in which various combinations of parameters can provide a wide range of optimized solutions. Both physical parameters and other attributes such as cost and environmental impact can be used in a single cost function.

Numerical optimization can provide solutions which are not intuitive as the method of solution is not guided by scientific theory. For this reason, optimization can yield results which lie outside commonly accepted design practice.

Exercises

5.1 Optimization review. Locate a published paper in which a device has been designed. Review the paper and note the parameter study to optimize the performance of the device. Has the device been optimized using a multi-parameter technique? Suggest what parameters need to be optimized simultaneously in order that the best possible result has been obtained.

5.2 Cost function. Locate a published paper in your engineering discipline in which optimization has been used. Write out the cost function and comment on the relevant weighting given. Note also the sensitivity of the optimized solution to small changes in the input parameters. Is such precision reasonable given the mechanical and electrical constraints of the system?

5.3 Environmental inputs to a cost function. Select and read a published paper in your engineering discipline in which some form of optimization has been used. Review the published solution and comment on other possible solutions in which the environmental effects have been considered. Do you now think that the published solution is optimal? Discuss your conclusion.

5.4 Simplex implementation. Use the simplex algorithm to solve for the location of the minimum value of x and y in the function $F(x, y) = y^2 + x^2 + 7x$. Use the point $(5, 7)$ as the start location.

5.5 Gradient optimization. Use the gradient algorithm to solve for the location of the minimum value of x and y in the function $F(x, y) = y^2 + x^2 + 7x$. Use the point $(5, 7)$ as the start location. Compare your speed of convergence with your result from Exercise 5.4.

References

Keywords: optimization, Monte Carlo, simplex method, gradient method, life cycle analysis, genetic algorithm, multiparameter optimization, pareto front

[1] Ciambrone, D.F., *Environmental Life Cycle Analysis*, Boca Raton: Lewis Pub., 1997.
[2] Stamm, A., James, D.A., Hagem, R.M. and Thiel, D.V., 'Investigating arm symmetry in swimming using inertial sensors,' IEEE *Sensors* Conference, Taipei, Taiwan, pp. 1–4, Oct. 2012.
[3] Sarkar, A.K., James, D.A., Busch, A.W. and Thiel, D.V., 'Triaxial accelerometer sensor trials for bat swing interpretation in cricket,' *Procedia Engineering*, 13, 232–237, 2011.
[4] Beale, E.M.L., *Introduction to Optimization*, New York: Wiley, 1988.

[5] Foulds, L.R., *Optimization Techniques: an Introduction*, New York: Springer Verlag, 1981.

[6] Gen, M. and Cheng, R., *Genetic Algorithms and Engineering Designs*, New York: Wiley, 1997.

[7] Rajeev, S. and Krishnamoorthy, C., 'Discrete optimization of structures using genetic algorithms,' *J. Struct. Eng.*, 118 (5), 1233–1250, 1992.

[8] Doringo, M. and Stutzle, T., *Ant Colony Optimization*, Cambridge MA: MIT Press, 2004.

[9] Poli, R., Kennedy, J. and Blackwell, T., 'Particle swarm optimization,' *Swarm Intelligence*, 1 (1), 33–57, 2007.

[10] Bertsimas, D. and Nohadani, O., 'Robust optimization with simulated annealing,' *J. Global Optimization*, 48 (2), 323–334, 2010.

[11] Thiel, D.V. and Smith, S. *Switched Parasitic Antennas for Cellular Communications*, Boston MA: Artech House, 2001.

[12] Koshel, R.J., 'Simplex optimization method for illumination design,' *Optics Letters*, 30 (6), 649–651, 2005.

[13] Krejic, N., 'A gradient method for unconstrained optimization in noisy environment,' *J. Applied Numerical Mathematics*, 70, 1–21, 2013.

6

Survey research methods

6.1 Why undertake a survey?

Engineering research only has value when it directly or indirectly contributes to the improvement of the human condition. One form of research is to seek information about and/or feedback from people about the outcomes or the proposed outcomes of the research, whether this is a product or a service. In addition, an assessment of the 'usability' of a product or service can be the subject of a research programme [1]. A failure to consult the potential users of developing technology may restrict its use in society. For example, the developer of new technology should recognize and address the limitations of that technology by addressing the following questions:

- Is the technology limited to a particular age group, ethnic group, or a group with disabilities – mental and physical?
- Will the technology cause unintentional harm to users and the environment?
- Is it possible that the technology can cause injury, disability and even death in the very worst cases?

One method of reviewing these questions is to seek feedback from the potential users and the public at large. In many countries, anti-discrimination laws restrict the design and use of community infrastructure which is not inclusive of sections of the population with disabilities.

There are also design challenges for devices that can cause potential harm, whether the device is used by the intended users

Example 6.1 **Disability challenges**

Is it possible to design doors that wheelchair users find convenient to use?

How is it possible to implement a fire alarm system for someone who is profoundly deaf?

Is the red, orange and green colour scheme used in traffic lights suitable for the 10% of the male population with red–green colour blindness?

Not only are creative designs required, the ability of these populations to use a new system must be reviewed through a survey.

Example 6.2 **Dangerous goods challenges**

Magnetic resonance imaging of the human body requires exposure to very intense static magnetic fields. This can have major detrimental effects on those with implanted heart pacemakers.

Children's toys must not be coated with lead-based paints. The toys must not have sharp edges or small parts that can be swallowed, etc.

The sale of portable high-powered lasers is restricted to scientific and manufacturing organizations. Members of the general public are not allowed to carry such items because of their potential danger to society.

The actions of mining can cause dust hazards, ground subsidence and pollution of local streams.

A survey might be used to establish the effectiveness of production and sale controls on these products and services. The result might be some restrictions on access and use.

or non-intended users who might have accidental exposure to the system.

The only method of obtaining some of this information in a rational, controlled manner from a representative sample population is through the use of a well-constructed, scientifically-based survey. There are two somewhat distinct engineering investigations that require feedback from user groups or the general population:

- Defining the characteristics of humans – physical, mechanical, chemical, genetic and physiological parameters including impairments. These are commonly described in terms of anthropometric, biomechanical and biochemical measurements.
- Defining the opinions of humans – what they think about objects, engineering plans and outcomes and the usability of objects and systems. These are commonly described in terms of social, cultural, religious and political aspects of human life. This can be described as a usability survey.

Generally, the first of these requires measurements of the physical and chemical parameters. In many cases, these anthropometric data are independent of time (at least in the short term) and so measurements can be made at any time. The exception is physiological measurements (respiration, blood chemistry, etc) where recent past exercise and food intake can significantly influence the results. For example, body shape (height, weight, etc) and motor control (reaction times, physical coordination and physical strength) change with age (see Figure 6.1). Studies have shown that the human body has changed with the generations – the result of improved nutrition (and excessive food intake in the case of obesity problems) and improved medical care. Also increased longevity is leading to an ageing population.

Responses to the latter question (based on opinion) can be time dependent and subject to influence from the media, immediate neighbours and workmates. Memories of the recent past can affect survey results significantly.

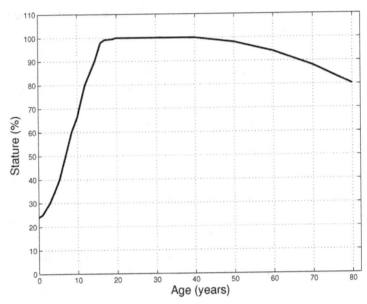

FIGURE 6.1 Change in height (stature) as a function of age [2]. The relative height is expressed as a percentage of maximum height during a lifetime. Height changes little between the ages of 18 and 50 years.

Example 6.3 **Anthropological characteristics**

Studies have demonstrated that the average size (see Figure 6.2) and weight of the population have increased significantly [3–5].

The average height of members of a men's basketball team is much larger than the average height of female gymnasts.

Should these factors influence the size of an airline seat in planes to be released onto the market in the year 2025?

In almost all cases, product development is best undertaken using concurrent engineering principles. That is, the technological aspects of product research and development are undertaken simultaneously with a survey used to determine the utility and applicability of the product or service. Researchers must therefore

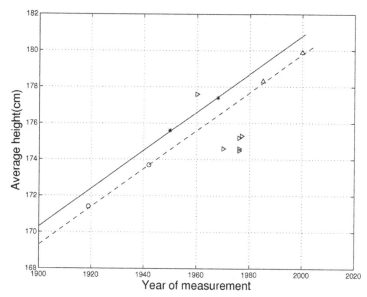

FIGURE 6.2 The increase in height for USA flying personnel (from [2]). The circles and the △ points are values for air force fliers and later NASA shuttle crews, The * and ▷ points are for air force flyers and army, marines and navy personnel respectively. The dashed line and continuous lines are linear fits to two sets of data, air force fliers and NASA crews, and air force personnel. The lines have been used as a predictive estimate of the future trend in height of these cohorts.

be aware of the human limitations in using and operating the technology and the basic requirements and principles underlying survey research methods. This includes reporting methods and the educational level, history and cultural aspects of the community.

Most surveys will have the following requirements:

- The population surveyed must be sufficiently large so that appropriate statistical analyses can be undertaken.
- The time allocated to complete the survey should be relatively short to ensure recent events do not impact on the survey outcomes.
- The population must be sufficiently large to ensure that individual identities cannot be inferred from the statistical data presented.

Example 6.4 **Public opinion**

A survey of highway safety conducted after a major accident is likely to yield different outcomes compared to a survey conducted when there have been no major accidents for a long time.

The general population continues to change its collective view of nuclear power in the light of nuclear accidents and global warming.

What is accepted as appropriate engineering technology and infrastructure in one community might not be acceptable in others.

To ensure technology is effectively and efficiently used, it is best to seek the opinions of the end users before spending money and resources developing and implementing new technologies.

- The population must be sufficiently diverse to ensure that population groupings are not subjected to adverse stereotyping.
- Participation is voluntary, and there is no mandatory requirement that all questions/sections of the survey are completed.
- The survey results are presented to the population surveyed in an understandable form.

Example 6.5 **Product optimization**

There is a compromise between the size of a mobile telephone and the need to provide a full character set keypad. The size of the soft keys on a keypad is of significant importance. If the keys are too small then the majority of the population can't use them effectively. Telephones must be usable by members of the population with disabilities including motor-neuron control problems, limited vision, unusual colour resolution, etc.

As an example, a level of human capacity (both physical and mental) might be required to ensure that the users of a particular product or service are capable of using it.

It is clear that making a product which is not usable by some sections of the population is discriminatory and can impact on the uptake of the technology. In addition, the capabilities of each individual are subject to the effects of their mental state and fatigue level and may be influenced by drugs and medication. A decision must be made about the range of activities required to use a particular product or service and this might be a design feature.

Example 6.6 **Levels of usability**

The option of disabling a motor vehicle when the driver is incapacitated by trauma, drugs or alcohol is a design feature in some vehicles. However, a mobile telephone should not have the same level of motor skills required. Thus a person under the influence of drugs or the onset of a medical condition should not drive a motor vehicle but should be able to use the telephone to call for a taxi or ambulance.

6.2 Ergonomics and human factors

The human engineering specification and the normal range of human functions is the subject of many research investigations in a field called *ergonomics*, and the measurements are termed anthropometric measurements. The field of research into the use of technology by humans is usually referred to as *human factors* research [6–8].

There are detailed lists of the proportions of limb segment length with respect to total height, etc [2].

Example 6.7 **Population measurements**

In a 1981 study the mean body height of USA civilians was 162.94 (6.36) cm for women and 175.58 (6.68) cm for men [2]. The numbers in brackets are the standard deviations about these mean values.

With improved medical and nutritional support, the population in most places in the world continues to live longer [9]. Older adult users have statistically different sensory function including taste and smell, movements, hearing, vision and different cognition (including memory, visual attention and spatial cognition) [9].

There are many textbooks [2, 6–9] and scientific journals which publish information about human capabilities – sensory perception (smell, vision, hearing, touch, taste), speed and range of limb movements, reaction times, concentration times, increased susceptibility to injury, etc. This type of information is vital to almost every engineering project, from transport systems (automotive cockpit design, highway design, door and chair designs, etc), to artificial heart value design, communications systems, building vibrations, etc. No two humans are exactly the same. Even identical twins are influenced through different past experiences and so will have different physical and mental characteristics.

The data on physical and mental attributes commonly follow a normal distribution. The standard deviation around the mean value for human attributes must be considered when designing engineering systems. Designing for the extremes allows maximum engagement for the whole population.

Example 6.8 Seating

An office chair should comfortably seat the shortest and the tallest individuals in the population for an 8 hour work period.

The seats in an automobile must be suitable for a wide range of adults, must be sufficiently restrictive to protect individuals in an impact, but sufficiently easy to enter and exit the seat. Some pregnant women, individuals with significant back pain or neck pain often have difficulty entering and leaving a motor vehicle.

A company designing chairs and seats must seek to accommodate the smallest through to the largest individuals in comfort in order to maximize sales and ease of use. Recent designs in office and motor vehicle seats provide a series of adjustments to accommodate large variations in the size and health of the users.

Rather than standard deviation, many population studies present their results in terms of percentiles [2]. Engineers must be aware of equity issues and inclusiveness for those in the population that have a disability – both mental and physical.

Example 6.9 **Stairs**

Public buildings designed in the nineteenth century featured large sweeping staircases at the front of the building. In the latter part of the twentieth century, staircases were hidden and elevators became common. At the beginning of the twenty-first century, with increased problems of obesity, architects are again delivering building designs where the staircases are obvious features. The use of stairs reduces energy use in the building. Now that architects provide wheelchair access to public buildings, so then both elevators and stairs are easily found features on the ground floor in the main entrance hall.

Many of these issues require feedback from the population. The role of a research team engaged in human factors research is somewhat different from the more standard experimental, observational and theoretic research projects. The research process can be summarized as [8, 10]:

- Finalize a research question;
- Develop the research plan;
- Design the survey instrument;
- Apply for ethics approval;
- Contact the sample population;
- Issue/distribute the survey (unless a third party is assigned this role);
- Prompt the sample population for responses;
- Analyse the data;
- Generate conclusions and reports.

6.3 Ethics approval

If the research team administers a survey, then there is a possibility of bias, and confidentiality can be compromised. Any suspicion of bias can compromise the research results. For this reason, many surveys are commissioned to be carried out by independent survey organizations. This is particularly true in organizations where management seeks to understand the thoughts of employees. Using third parties to remove bias and maintain confidentiality minimizes this risk. There are commercial organizations that will conduct surveys and there is survey software available which prevents the identity of individuals being disclosed to the survey team.

Example 6.10 Bias and confidentiality

Clearly it is difficult for workers to comment on their working conditions or their pay when directly confronted by the boss.

It is difficult for a student in a class to comment on the quality of the classroom teacher.

The survey outcomes are likely to be biased as the person asking the questions has power and influence over the individuals who are the respondents. Even if the responses are not biased, the mechanism is prejudicial and the results cannot be relied upon.

As a survey involves seeking the opinions or personal details of others, ethics approval to conduct the research is mandatory. Most learned journals will not allow the publication of survey results unless the appropriate ethics approval number from a certified ethics committee is cited in the paper. The requirement of ethics approval is to ensure that the cohort of people whose opinions are sought, are not inadvertently damaged (either physically, mentally or reputationally).

The application for ethics approval commonly requires a justification for the research (what is the aim of the research and why is it important?), and the submission of three documents:

- Information sheet. (This describes the project aims, methods, requirements of the participants, personal details information, and includes the names and contact details of the researchers.) This sheet is passed to the participants before the measurements begin. Participants are encouraged to keep the sheet.
- Consent form. Here the participants sign and date the sheet. The sheet is retained by the research team. Each participant is assigned a number.
- Survey sheet. This is the sheet which is completed by the research team and/or each participant. Only the participant number is written on this sheet. The name of the participant is not included. This allows the information to be stored separately from the consent form.

Most research organizations have a research ethics committee with members who are trained in medicine, law, psychology and research practice. In some countries every ethics committee must be registered by a government authority who will ensure that appropriate skills and training are available within their membership. A research team that has received ethics approval might be audited during or after the research has been completed. An audit is designed to ensure the research was conducted responsibly, that all participants have signed consent forms and that the research outcomes have been made available to the participants.

6.4 General survey guidelines

There are many textbooks which describe strategic approaches to survey design [11, 12], and there are semi-automated tools available on the web (e.g. Google survey tool, SurveyMonkey, etc.). The general rules are summarised in the following sections.

Rather than asking questions, surveys should seek the reaction of the respondents to carefully worded phrases or sentences. Survey instruments (questionnaires) consist of several sections (see Figure 6.3).

Section 1: Introduction

This is an explanation of the reasons for the survey, assurances that the responses are anonymous and details of the time and place where survey results will be published. Contact details and ethics approval number must also be given.

Section 2: Instructions

These are instructions to the participants about how to complete the survey. For paper surveys, respondents might not provide all of the information sought (i.e. some respondents will leave some statements without a score). In web-based surveys the researcher can ensure that all sections are completed by making it impossible to progress through the survey until every statement in every section has a response.

Survey title				
Introduction (Outline the reasons for the survey, the survey team, the ethics approval number, research team contact details, time lines etc)				
Instructions: Circle/highlight one response to each of the following statements				
Survey Statements Strongly disagree				Strongly agree
Statement 1	1	2	3	4	5
Statement 2	1	2	3	4	5
Statement 3	1	2	3	4	5
Statement 4	1	2	3	4	5
Statement 5	1	2	3	4	5
More statements				

Demographics:

My age is	< 18 years	18–25 years	26–40 years	46–65 years	> 65 years
My highest level of education is	Primary school	Secondary school	Apprenticeship	Uni degree	Uni higher degree (Masters/PhD)
I have been employed full time for	< 1 year	1–5 years	6–10 years	11–15 years	> 15 years
My house/flat has ? adult residents	1	2	3	4	> 4
I live ? kms from my place of work	< 3	3–10	11–20	21–50	> 50
More statements					

Open statements:
1>   What do you consider the best .....

2>   What do you consider the worst ....

Thank you for your participation. A copy of the report will be available on the web site http://xxx

FIGURE 6.3 The major components of a survey instrument.

Section 3: Statements

Respondents should rank their answers to the survey statements according to a 4, 5, 7 (called the Likert) or 10 point scale. The scales are usually outlined as follows:

Clearly the 4 point scale and the 10 point scale do not allow the respondents to return a neutral view. Either they agree or

4 point: *Strongly disagree   Disagree   Agree   Strongly agree*

5 point: *Strongly disagree   Disagree   Neutral   Agree   Strongly agree*

7 point: *Strongly disagree   1 2 3 4 5 6 7   Strongly agree*

10 point: *Strongly disagree   1 2 3 4 5 6 7 8 9 10   Strongly agree*

disagree. The 5 point and 7 point scales do allow a neutral or undecided answer. For some statements it might be appropriate for the respondents to make no comment rather than commit to a view which will bias the result. This requires an additional response column for *not applicable/no opinion/don't know* responses.

Section 4: Demographics

Respondent profiling statements are used to characterise the responses of different sub-groups in the population. For example, age, height, weight, education, place of birth, years of employment, gender, etc might all be possible reasons for a particular view to be expressed in the survey responses. Note that for anonymity and ease of data processing, it is best to characterize the responses to these questions using ranges aligned with the same numerical scale used in the statements (Section 3).

For example, the demographic statements to be selected at the end of a 5 point survey might read:

Statement	Strongly disagree	Disagree	Neutral	Agree	Strongly agree
Survey statement 1	1	2	3	4	5
*					
*					
*survey statements *					
Survey statement 25					
My age is	< 18 years	18–25 years	26–40 years	46–65 years	> 65 years
My highest level of education is	Primary school	Secondary school	Apprenticeship	University degree	University higher degree (Masters/PhD)
I have been employed full time for	< 1 year	1–5 years	6–10 years	11–15 years	> 15 years

Section 5: Open ended questions and comments

The research team can learn more about the thoughts, background and motivation by inviting responses to open ended questions and comments. Assuming *xxxx* is the topic of the survey, the open questions and comments might include:

List the best attributes of xxxx.

List the worst attributes of xxxx.

Give one or more examples of a positive experience using xxxx.

Give one or more examples of a negative experience using xxxx.

What needs to be done to make xxxx more widely acceptable?

When reporting the research outcomes it can be useful to include some exact quotations from insightful comments from one or two respondents which support the overall conclusions. This is particularly true in relatively short surveys where the issues cannot be explored in great depth.

6.5 Survey statements

The construction of survey statements is critical to achieving reliable outcomes from a survey. There are a number of general guidelines which should be followed [11, 12].

6.5.1 Statements should all be positive

This allows the respondent to complete the survey quickly if they wish. They do not have to read each statement carefully. If they generally agree with the concepts presented then they do not have to move their cursor or pen much between the statements to register a strongly positive or strongly negative response.

There is nothing to be gained by making the statements highly complex or varying the statements between positive and negative. This latter approach generally annoys respondents with the result that some bias might creep into their response. They could accidentally miss the negative in the sentence and so record a vote which is diametrically opposed to the view they wish to express.

Example 6.11 Refining your statements

This is a poor statement: *I don't feel well most of the time* (negative wording).

This is a better statement: *I feel well most of the time.*

6.5.2 **Statements must be short**

Surveys must contain concise statements without lengthy explanations. In addition the number of questions must be small. Commonly survey designers should aim to restrict the survey to a maximum of 20 items presented for comment. Surveys are more effective when they can be completed by most people in less than 5 minutes.

There is a significant temptation for survey designers to ask a very large number of questions. While minimizing the risk of omitting important information which might be useful in addressing the research question and other, as yet, undefined research questions, the survey designer must consider the quality of the responses gained from a long, complex survey. As ethics approval is only given for a specific project aimed at answering a particular research question, it is unethical to seek more information about participants than is needed to answer this specific research question. Commonly ethics approval is given only after the survey instrument has been prepared and examined by the committee. Ethics approval will only be given if the researchers can defend the relevance of the information being sought. For this reason, no extraneous questions must be included in the survey.

Most people are subjected to surveys from a number of different sources, for example professional survey companies seek information from the general public about food products, political opinions, consumer goods, etc. This can be annoying and many people refuse to participate. One of the ethical requirements in scientific surveys is that the participants must be able to obtain a summary of the survey findings when the research project (which includes the survey) is complete. Many surveys indicate at the start of the survey, how long the research project is likely to take to complete. This can influence a person's decision to participate at the time when he/she receives the survey.

Frustration of the users is a major impediment to gaining reliable data from surveys. A frustrated participant might choose a negative answer for all questions regardless of their true opinion.

Some might chose a neutral answer for all questions. It might be appropriate to remove such survey returns from the data analysis.

Long surveys can add to the frustration. If the potential respondents are forewarned about the expected time for completion, they are less likely to become frustrated or annoyed by the survey.

6.5.3 Statements must use clear language

The statements in a survey should not be complex. The use of acronyms and idiomatic phrases and slang words must be avoided. The respondents might not understand the statements and consequently give misleading responses. The use of double negatives and unusual phrasing must be avoided.

Example 6.12 Survey statement examples

This is a poor statement: *Hydraulic oil should not be used as a lubricant in electric motors unless well insulated.*

This is a better statement: *Provided that the motor wiring is well insulated, hydraulic oil is a good lubricant in electric motors.*

This is a poor statement: *The EM radiation scattered by an RFID antenna is a measure of efficiency.*

This is a better statement: *An efficient antenna will strongly scatter electromagnetic waves.*

This is a poor statement: *The University does not do a bad job in keeping us informed about workplace health and safety issues* (double negative).

This is a better statement: *The University does a good job in keeping us informed about workplace health and safety issues.*

6.5.4 Statements should not attempt to trick the respondents

Survey designers should not seek the same information more than once using a number of statements which contain different phrases or negative and positive statements to 'check' the authenticity of the responses. Trying to trick the respondents has a number of negative effects:

- It can frustrate the participant. This can lead to poor quality feedback from the survey.
- It makes data analysis extremely difficult. If the answers are contradictory, which answer do you chose? If there is a contradiction it might be necessary to discard the survey responses to this statement.

6.5.5 Statements should not have multiple themes

If two concepts are used in the same statement then it is impossible for the respondent to choose a single response.

> **Example 6.13 Double statements**
>
> How would you score the following statement?
>
> *I am satisfied with the ventilation and the lighting environment.*
>
> If there is a problem with the lighting but not the ventilation, then the respondent cannot convey this information by a single score.
> A better set of statements is:
>
> *I am satisfied with the ventilation in the building.*
>
> *I am satisfied with the lighting in the building.*

6.5.6 Statements should not direct your feelings onto the respondents

It is a relatively simple matter to direct the survey statements in such a manner that the responses that are sought or expected are returned by the participants. The use of emotive words and descriptions must be avoided.

Example 6.14 Directing the respondent's feelings using emotive language

This is a poor statement:

Many people feel ill as soon as they walk into the building.
This statement projects the feelings of the research team.

This is a better statement:

People enjoy working in this building.

6.5.7 Comparative statements should not be used

Without a knowledge of the past experience of others, the survey might not determine absolute views from the respondents if comparative statements are used.

Example 6.15 Comparison statements

This is a poor statement:

This work environment is just as good as other places where I have worked. (This is a comparison statement. A person who has not worked elsewhere cannot respond.)

A better statement is

I am happy with this work environment.

6.6 Survey delivery

A number of different methods can be used to distribute the survey. All have advantages and disadvantages.

6.6.1 Paper surveys

A paper survey can be distributed using mail (post or email) and the respondent is asked to complete the survey and return using the same mechanism – prepaid envelope or scanned *.pdf document attached to an email.

In order to participate, the respondents must have a mail or email address. Some sections of the population have neither (either by choice or disadvantage).

Example 6.16 Respondents with disadvantage

A homeless person or a transient worker is unlikely to have a postal address that is used regularly.

A large percentage of elderly people do not have access to a computer and so will not have email services.

People with poor skills in the language of the survey, nomadic groups and illiterate people cannot be contacted reliably in this manner.

It is clearly impossible for a person without a mail address to answer questions about the quality of the mail delivery, of a person without a car to comment on traffic delays, etc.

A member of the research team might travel to a particular location with a clipboard and the survey sheet (shopping centre? coffee shop? library? etc). The team member would invite passers-by to answer the questions by oral questioning and recording the responses on the score sheet.

The tone of voice used by the questioner and the physical appearance of questioner can influence the responses. The location of the questioner is also prejudicial to the respondent status.

Example 6.17 Shopping centre surveys

Shopping centres commonly attract the following types of people:

- People with money to spend;
- People seeking shelter (from the rain, for warmth, to be cool, etc.).

Shopping centres do not attract those without money and those who engage others do their shopping for them.

The time at which the questioner is there will influence the number of full-time employees who complete the survey because of shift work, and the number of those with parental or carer responsibilities, etc.

6.6.2 Telephone surveys

Through random selection of contact numbers from the telephone book/directory, the questioner rings numbers and seeks responses. The questioner records the responses on a score sheet or computer.

The population of respondents is limited to those who can be contacted by publicly available telephone numbers at the time of day when they are available.

6.6.3 **On-line surveys**

Through random selection of email addresses or using social media, the research team can solicit survey responses. The bias in this method lies in the population who regularly use these methods of communication.

6.7 Respondent selection

As with all engineering research projects, the research team needs to identify the stakeholders. Respondents will participate in a survey for one or more of the following reasons:

(a) Public good: The respondent sees the outcomes of this research benefiting the community, the nation and the world. Sometimes described as the 'warm inner glow', there is no personal benefit in engaging with this research. The respondent is simply seeking to help human kind.

(b) Personal good: The respondent sees the outcomes of this research benefiting his/her world view of how the scientific community and governments should progress the matter under discussion. While there is no immediate, tangible reward gained, the respondent seeks to influence public opinion which might benefit him/her later.

(c) Personal reward 1: All respondents automatically and anonymously are given a gift – for example a redeemable voucher for an on-line store product or a music download.

(d) Personal reward 2: The names of the respondents are entered into a prize draw or lottery and a small number of respondents are selected at random to receive this reward.

It is clear that in the first three cases anonymity can be preserved. The respondent is simply allocated an identification number for inclusion in the spreadsheet. In the last case, the name and contact details of the respondent must be included in the response and anonymity is compromised. For this reason, the ethics approval application must clearly state if and how the respondents will be rewarded. The ethics approval committee needs to approve any incentives offered to participants.

Following the identification of the potential respondents, it is necessary to review the likely biases that might be present in the responses from the different sub-populations. The aim of using a survey is to identify the opinions of various population groups. While every individual will have a slightly different perspective on the research question, care must be taken to ensure that there is not one sub-population that dominates the research conclusions. This problem cannot be avoided completely, but an awareness of bias can temper results. Usually the questions in the demographics section of the survey are selected to ensure these different population groups are identified. The differences between two groups of respondents can be assessed statistically using an unpaired t test.

In public discussions, there are some people who are highly offended if their views are not sought.

Example 6.18 Interested groups of respondents

The people that live close to proposed new infrastructure developments desperately seek to voice an opinion.

Some religious groups seek a voice in new biomedical technologies.

The owners of apartment buildings seek to influence the move to a six-star energy rating for premises, as it introduces additional costs.

The road safety councils have a responsibility to comment on proposed new road safety guidelines and smart highways.

When selecting the respondents, attention must be given to engaging the relevant interested groups.

If the survey responses are anonymous, then the sample population will be less influenced by the person/organization asking the questions. The respondents might also be less concerned about others learning of their personal opinions. If the survey seeks to reward the respondents with a gift, then it is important that the names and addresses of the respondents should not be stored in the same place as the survey data.

There might be a research reason for a follow-up survey seeking further clarification on some themes, or if the research objective is to record the views of respondents before and after an event, then the list of contacts will need to be stored. Again, the location of the store of names should not be the same as the store of the response data. These issues must be addressed in the ethics approval process. The statistical analysis of before and after surveys should be conducted using a paired t test.

Longitudinal surveys seek to track the characteristics of populations over periods of time. In some surveys inter-generational attributes are tracked. In many cases it is necessary to track the changes in each individual separately. This means that, while the survey information must remain anonymous to the research team, the respondents must be able to identify themselves with their previous survey responses. This can be accomplished through the use of an identification number allocated to the respondent and stored independently of the name. The paired t test should be used for statistical analysis of two measurements on the same population, and ANOVA analysis for more than two surveys of the same population.

6.8 Survey timelines

Once the research question has been decided and the research methodology includes a survey (often called a survey instrument), the population of respondents must be determined, the instrument wording is finalised and then ethics approval must be sought. The ethics approval committee needs to understand the reasons for the research, the method of participant selection and the wording of the survey – both the instrument and the purpose of the research.

The survey must then proceed within quite tightly defined timelines to ensure the motivation for participation in the survey is not forgotten by the potential respondents. A typical timeline for a survey is given in Figure 6.4.

It is essential that the preliminary report is not released publicly before the closing date of the responses. There is a possibility that knowledge of the preliminary results can influence the responses of later respondents.

In the event that the number of responses is insufficient for statistical analysis (for example if the research needs respondents from different subsections of the population, e.g. age or location of residence, etc) and representation from each group is not sufficient in the first survey, it may be necessary to conduct a new survey. Surveys conducted by telephone calling often cease the interview if the respondent is in a group in which the requisite number of responses has been received. This strategy can be counterproductive if the sub-populations are treated independently. The larger the number of respondents, the better is

Week Number	Activity
0	Ethics approval granted
1	Pre-survey letter of introduction which outlines the research context and the research plan
2	Survey is sent out to the potential respondents
3	Mid-survey reminder letter
4	Final, last minute reminder to potential respondents
6	Post-survey preliminary report is made a available to all respondents
6	Thank you note to respondents with major conclusions

FIGURE 6.4 Sample timelines for survey completion.

the statistical representation of the true population responses. If, however, the representation is not evenly distributed across each sub-population, then one of the sub-populations can dominate the results and skew the outcomes. These considerations must be discussed in the planning stages before the survey is commenced. The analysis of results must take this imbalance into account.

Example 6.19 **Telephone survey bins**

In undertaking a telephone survey using the available fixed line telephones that can be publicly accessed, people less than 30 years old generally do not have permanent accommodation (they move from flat to flat in a 6 month or yearly cycle) and only use their mobile telephone. It is too expensive or too complicated to continue to arrange for the allocation and installation of a fixed line telephone. This means that the survey is likely to be skewed towards respondents who have purchased a house and so have permanent employment. The response rate of minority groups can be increased if these groups are specifically targeted using an alternative strategy.

6.9 Statistical analysis

A 5 point survey with 50 respondents and 25 statements (including demographics) results in a 25 × 50 element matrix. All of the responses are converted to numbers. For example a single respondent to a list of five statements will contribute one row to the matrix and the row contains integers in the range between 1 and 5. It then becomes a relatively simple matter to generate response histograms for each question and to undertake correlation analysis between different columns (i.e. the responses to individual statements). These types of statistical analysis and methods of presenting the results are described in Chapter 4.

Multi-dimensional correlation analysis can also be undertaken to minimize the effect of a third question result or demographic divisions when assessing the correlation between two other responses.

It might be of interest to characterize the responses of sub-populations based on their demographic responses. In this case the total population of respondents must be divided into groups on the basis of the demographic responses. One can then use the unpaired t test to compare the population means and standard deviations and to determine the probability that the populations are different.

When individuals are surveyed twice (i.e. before an event and after an event) the paired Student's t test can be used to calculate the probability that there has been a change in the responses. When two different populations are surveyed or the population can be divided into two or more different sub-populations, the unpaired t test can be used to test the probability that there are differences between the two populations.

6.10 **Reporting**

The usual scientific style of reporting is required to define the research outcomes for publication in the research literature. However, there is an obligation on the research team to provide feedback about the research outcomes to the respondents. Commonly this can be done via the web or direct email communications.

As a scientific report might be difficult to understand for many people, it is necessary to ensure that the report can be understood by all respondents. Thus the major findings must be reported in a journalistic style usually found in newspapers. Anecdotal evidence suggests that this style of writing is understandable for those with a grade-10 level of education. Most newspapers and television news reports include histograms and 2D graphs to illustrate statistical information and so this reporting technique should be used in the report provided to the respondents.

One common method of reporting is to calculate the percentage of respondents who have expressed a positive view on one or more of the survey statements. Using a scoring system based on an even-number scale mandates that the respondents express a positive or negative opinion and cannot remain neutral. The removal of the neutral score option can increase the strength of conclusion statements. This might not be a completely valid finding for those respondents who really preferred to remain uncommitted to expressing an opinion.

When addressing the concerns of a minority section of the population, the conclusion statement might be '20% of the

Example 6.20 **Survey statistics**

Let us assume that agreement responses (strongly agree or agree) for one survey statement totalled 40% with 30% of respondents selecting neutral. The conclusion from this survey can be written as: 40% of respondents agreed with the statement.

In an even scale, the 30% would be forced to score agree or disagree. Assuming an equal split in the neutral vote, the conclusion becomes more convincing: 55% of respondents agreed with the statement.

The research team needs to make a decision about which result is more appropriate and accurately reflects the views of the respondents.

respondents did not agree' with the statement. This tends to be more powerful than quoting the 80% positive responses.

The reporting of open questions can be undertaken in a number of different ways. Positive statements can be counted and used (via quotes) to further support the numerical data. Selected, representative quotes can be used to highlight a depth of feeling which is not obvious from the numerical scores.

Example 6.21 **Summary statements of results**

One might summarise findings from a nursing home survey by a statement like: More than 80% of the respondents agreed that face recognition software in private rooms was a good idea, and more than 20 open comments were received with statements like 'This innovation would dramatically improve the condition of those suffering from dementia'.

6.11 Chapter summary

The work of engineers is designed to improve the human condition. This requires an understanding of the human size, shape, movement degrees of freedom, mental and physical capacity and sensory systems. This information can be obtained through testing and surveys. Researchers must recognise that the characteristics of human beings continue to change with time, firstly as people age, and secondly, as a result of generational change predominantly caused by improved medical services and nutrition. As people are living longer than previous generations, a new set of engineering requirements must be developed so that the elderly can be accommodated in normal society. This includes addressing such issues as dementia and muscular–skeletal degradation [9]. This is an active field of engineering research as the average age of the population continues to increase.

It is important that various populations of people are given the opportunity to express an opinion about the directions of engineering research. This should ensure that the research team is involved in the development of products and services that are acceptable to the target population of users and the local community.

The most rigorous method of gaining an insight into the opinions of the target users and the general population is to create and implement a survey. As humans are involved in this research, so then ethics approval must be gained before the survey is undertaken.

The survey instrument must be structured as a series of statements and the respondents are asked to comment on those statements by expressing various shades of agreement or disagreement. The research team must take care to ensure the reliability of responses through using clear statements, allowing only a short time to complete and ensuring that it selects an unbiased population of respondents.

The analysis of the data can be undertaken using standard statistical techniques including histograms, multi-dimensional correlation and t test probability calculations. Longitudinal surveys require a form of respondent identification while the ethical requirements mandate that the data must be stored with identifiers without the respondent names.

When a research group is contemplating the development of a survey, they might review some exemplar surveys available in books and the published literature [13, 14].

Exercises

6.1 Research question. 'Does working in this building make its occupants feel sick?' Use a team of researchers to design a 25 statement survey which seeks information about the building in which they work. Ask different individuals to write three statements each covering the following aspects: lighting, noise, smell, access, and safety and security. Another individual should be responsible for the project outline and respondent instructions and the open questions at the end of the survey. The research team should review the submissions from each individual and test the validity of the statements by answering the statements individually. The survey should be revised until all members of the team are satisfied.

6.2 Following the survey designed in Exercise 1, implement the survey on a group of building users and analyse the results. Answer the following questions: Are the results dependent on age? gender? educational status? The team should postulate reasons for the results.

6.3 Apply the knowledge gained from reading this chapter to a survey you have been asked to complete. Check which guidelines have been followed and which guidelines have not been followed.

6.4 Add the word 'survey' to your discipline keywords in a scientific search engine and check the search findings. Read one or more of the papers and check the method used for the selection of respondents and the method of reporting used in the paper. Write an 800 word summary of the research methods and outcomes from this research paper.

6.5 Check the scientific literature (textbooks and journals) for the mean and standard deviation of height and weight for adults in your community. What effect should this have on (a) the size of the seats on public transport? (b) the ceiling height of public buildings?

6.6 Check the scientific literature (textbooks and journals) for the mean and standard deviation of oxygen consumption of fit adult humans who will be involved in a spacecraft or underwater employment. From this information calculate the volume of air required per person per minute.

6.7 Check the scientific literature (textbooks and journals) for the mean and standard deviation of running speed of children aged between 5–10 years. From this information estimate how long it will take to evacuate a primary school class of 20 students through a 70 m long corridor.

References

Keywords: survey, ergonomics, human engineering, human factors, anthropometric, human factors, questionnaire, qualitative research methods, usability assessment

[1] Sohaib, O. and Khan, K., 'Integrating usability engineering and agile software development: A literature review', Int. Conf. *Computer Design and Applications*, 2, 32–38, 2010.

[2] Kroemer, K., Kroemer, H. and Kroemer-Elbert, K., *Ergonomics: How to Design for Ease and Efficiency*, Upper Saddle River, NJ: Prentice Hall, 1994.

[3] Meredith, H.V., 'Findings from Asia, Australia, Europe, and North America on secular change in mean height of children, youths and young adults', *American Journal of Physical Anthropology*, 44(2), 315–325, 1976.

[4] Bock, R.D. and Sykes, R.C., 'Evidence for counting secular increase in height within families in the United States', *American Journal of Human Biology*, 1(2), 143–148, 1989.

[5] Chike-Obi, U., David, R.J., Coutinho, R. and Wu, S-Y., 'Birth weight has increased over a generation', *American Journal of Epidemiology*, 144(6), 563–569, 1996.

[6] Sandom, C. and Harvey, R.S., (eds.) *Human Factors for Engineers*, London, UK: IEE, 2004.

[7] Phillips, C.A., *Human Factors Engineering*, New York: Wiley, 2000.

[8] Nemeth, C.P., *Human Factors Methods for Design: Making Systems Human-Centred*. Boca Raton: CRC Press, 2004.

[9] Fisk, A.D., Rogers, W.A., Charness, N., Czaja, S.J. and Sharit, J., *Designing for Older Adults: Principles and Creative Human Factors Approach*, Boca Raton: CRC Press, 2004.

[10] Meister, D., *Conceptual Foundations of Human Factors Measurement*, Mahwah, NJ: LEA Pub. Co, 2004.

[11] Mitchell, R.C. and Carson, R.T., *Using Surveys to Value Public Goods: The Contingent Valuation Method*, Washington USA: Johns Hopkins University Press, 1993.

[12] Creswell, J.W., *Research Design: Qualitative, Quantitative, and Mixed Method Approaches*, 3rd edition, Los Angeles: SAGE Press, 2009.

[13] Lewis, J.R., 'IBM computer usability satisfaction questionaires: psychometric evaluation and instructions for use', *International Journal of Human-Computer Interactions*, 7(1), 57–78.

[14] Nardi, P., *Doing Survey Research*, 3rd edition, Boulder, Co: Paradigm, 2013.

7

Research
presentation

7.1 **Introduction**

Assume that the process of research is approaching a conclusion. The research team has written the proposal, undertaken a critical review of the literature, put together the research question and the research team, the research results are available and have been analysed. But the research is not finished. Following previous discussions, the research is not complete until the results have been peer reviewed and published. This chapter describes the process of presenting research results for publication and for presentation in scientific and engineering research conferences.

The good news for the research team is that some of the work required for publication and presentation should be almost complete, as the background to the problem, the literature review and the research methods should have been completed when a research application was submitted. In addition the research application might have included some suggestions about how the results will be displayed graphically and what statistical methods were planned. Before writing the final report it is wise to check the published literature to see if there have been recent publications in the field. This chapter is about presenting the results in a concise, clear manner. The general format of a paper, thesis, abstract, presentation, etc should follow the same general form as illustrated in Figure 7.1.

The first part of a publication or presentation needs to be directed at an educated lay audience. That is, the researchers must

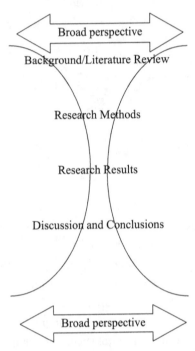

FIGURE 7.1 Structural outline of a scientific or engineering paper illustrating the change from a broad perspective to a narrow description of the experimental work conducted and its interpretation, and then concluding with a broad perspective to explain how the results impact on the current body of knowledge. This template should be used for all publications including journal papers, theses, abstracts, presentations, posters and even undergraduate reports of laboratory experiments.

place their research contributions in the wider context by addressing the question

How will this research benefit humanity?

The final section of the paper should also address the wider humanitarian context by answering the question

How can the results be implemented so that humanity will benefit from this innovation?

Between the beginning and the end of the research report, the research team must reveal the research and analysis methods upon which their conclusions are based. In so doing, the research team and presenters must provide an accurate description of the research and the raw data so that a reader of the paper or a listener to the presentation will have sufficient information to develop the same conclusions as reported by the research team. In full length papers, other research teams located anywhere in the world should be able to read the paper and gain sufficient detailed information about the experimental methods and procedures so that they could independently undertake similar experiments which would result in data from which the same conclusions can be drawn. A failure to do so has a number of consequences:

(a) The research will not be accepted for publication/presentation on the grounds that the information in the paper is not novel or relevant.

(b) The research team will be subject to questioning after publication or during the presentation about their research methods and the analysis of results.

(c) Another research team which attempts to repeat the published work, or bases further work on these results, might find flaws in the experimental method and/or the conclusions drawn. This will cast doubt on the validity of the published research and the capabilities of the research team. This has implications for the reputation of the team – particularly if the research is later revealed to be significantly flawed.

Research teams seeking publication and presentation at conferences must therefore engage in the review process in a positive light. The peer review process can add merit to the paper by strengthening the arguments, improving the logic and clarifying the background literature. Of course it is disappointing for a research team to have a paper rejected, but the major motivation of reviewers must be to ensure that papers which are published or presented are understandable in the defined context, and are valid in every sense. For all researchers (and in particular novice researchers), the reviewers' reports constitute expert opinion at no cost. Not only that, the research team has the opportunity to gain the attention of eminent researchers in the field if the paper is very strong and noteworthy. This can only benefit the reputation of the research team. Clearly the submission of poor quality research and badly written papers will have the opposite effect – the reputation of the research team will be degraded.

For these reasons, and for the public good, reporting research outputs requires significant time, clarity of purpose and most importantly, many critical revisions before a paper in preparation is submitted to a journal editorial board or technical programme committee of a conference. In this chapter some simple guidelines are given to maximize the chances of a paper being accepted.

7.2 **Standard terms**

It is very important that the paper/presentation can be understood by the reading and listening audience, and it must be prepared accordingly. Commonly, the audience can be assumed to consist of students, professors and research workers in the field or discipline. For this reason, the presentation must use terminology that is well known to the audience and appropriate to their level of education and disciplinary background.

A first simple rule is to use terms that are commonly found and used in the textbooks offered to first and second year university level engineering students around the world.

A second simple rule is to use research and statistical methods that are defined in textbooks, standards and dictionaries relevant to the discipline. The authors should not invent/create new terms and should not name equations after themselves or their research organization. It is up to other independent researchers in the field or standards committees to refer to names of the creators of new engineering ideas.

Example 7.1 **Naming conventions**

It is highly unlikely that James Clerk Maxwell called his equations 'Maxwell's equations' and Newton did not define a 'Newtonian fluid' or 'Newton's first and second laws'.

A third simple rule is to use symbols in equations that are consistent with textbooks and other papers. All symbols in an equation must be identified in the text immediately before or immediately after the first equation that uses them.

Example 7.2 **Defining symbols**

Equations can be introduced in sentences as follows:

'The voltage V can be determined from Ohm's law

$$V = \frac{I}{R_L},$$

where I is the current through the component and R_L is the resistance of the component L.'

Note that the symbols in the text must be identical in size and font to those used in the equation unless superscripts and subscripts are used in the equation.

The names of commercial products should not be used in the main text, and almost all acronyms (a 'word' formed from the significant letters of a group of words) should be defined as a set of initials immediately following the first mention in the text.

Example 7.3 **Acronyms**

The word lidar (light radar) is quite acceptable but the name of a particular lidar unit should not be used. Commonly, if the lidar unit used in the research is a commercial unit, then the name of this unit should only be found in the list of references at the end of the article. An example of the way to write such sentences is as follows:

'The distance to the object was determined using lidar [xx].'

Here [xx] is the reference to the commercial name of the unit and its details in the reference list.

7.3 Standard research methods and experimental techniques

There is a significant advantage in using experimental and computational techniques that have been previously verified in textbooks, reference books, and other refereed publications. This reduces the requirement for a detailed description of the experimental method in the paper. Only the result need be quoted if there are references given to these previously published works. There is still a requirement to specify the operational conditions and any unique aspects of the research methods used.

Example 7.4 Novel applications of known techniques

One might write:

'Custom designed finite element (FEM) code was developed using a novel boundary condition at the interface between the oceans at great depth. This FEM code was outlined and validated by Sajad [xx]. This new boundary condition used mesh elements described as...'

One might write:

'The hardness of the surface of the material was assessed using the cone penetration test [xx] with a 1 kg cone (tip radius 0.003 m) released from a height of 0.5 m above the sample.'

7.4 Paper title and keywords

The research team must write a title for their research paper which is unique and sufficiently descriptive to inform the readership of the research direction and field. Commonly this will require ten or more words and some specific information relating to the field of research.

The keywords included in the paper are important for a number of reasons. Firstly, they will be used by web-based search engines so that this new paper will be readily found by the appropriate readership. Secondly, the keywords are often used to select reviewers. Inappropriate keywords will result in the paper being sent to reviewers who might not be experts in the field. This will disadvantage the reviewing process and might result in:

- The possible rejection of the paper because the reviewers are not experts in this specific field of research;
- The acceptance of the paper with major flaws that might have to be retracted later; or
- Lengthy delays as more and more nominated referees refuse to review the paper on the basis that it is outside their research field.

Note that usually an Associate Editor nominates reviewers for every paper. This is a huge task which is often undertaken on a voluntary basis. The Associate Editor and reviewers are put under some pressure to make the review decision swiftly so that the time between submission and a decision (accept, revise or reject) is minimal. A careful selection of keywords is an important aspect to securing a positive and timely outcome to the review process.

Example 7.5 **Keywords**

The research team has completed writing the paper. From the references listed in the paper, seek out the keywords used in those papers. From these keywords, select appropriate keywords for the new paper.

7.5 **Writing an abstract**

When used in the scientific literature the term 'Abstract' refers to a brief, but total, summary of the research reported. It is common for journal papers to include an abstract so that the potential reader gains a brief outline of the research project, methods and outcomes. For conference papers, the abstract of the paper might be all that is reviewed before acceptance or rejection of the paper. Commonly the title, authors and abstracts are freely available using the academic search engines and the internet. Some journals (called 'open-access' journals) provide a full copy of the paper, but others will only provide a copy of the full paper for a fee. For these reasons the abstract is a very important part of the paper and needs to be carefully worded so that reading the title and abstract alone is sufficient to convey understanding of the research and its outcomes. Should this be of sufficient interest, then the full paper will be consulted.

In many cases there are special restrictions placed on an abstract. For example the number of words or characters might be limited, no references are allowed, no symbols are allowed, etc. Regardless of these restrictions, the abstract should follow the structure shown in Figure 7.1.

Assuming the maximum allowable word count is 100 words and assuming a maximum of 20 words in each sentence, then the abstract is limited to six sentences or less. Following the format given in Figure 7.1, the abstract should be written using the following form:

Example 7.6 **Abstract writing example**

Consider a news report of a football match:

'The world cup brings the best football teams of the world together in one tournament. The match between Iraq and France was a first round match in the FIFA World Cup. The match was played in San Paolo in heavy rain and strong winds. The two teams were evenly matched and both played open fast football with the ball crossing the centre line more than 30 times in each half. There were 21 shots on goal. The result has clearly established the strength of the teams and had a significant bearing on the World Cup outcome.'

Clearly it says what happened but there is one very important thing missing. This is the score – the major result. Every abstract must have a result of major importance to the reader – usually this includes a numerical 'score'. That is, a measurement of major significance. The insertion of the score makes the report clear and will satisfy the reader. This report should be written with the inclusion of results.

'The world cup brings the best football teams of the world together in one tournament. The match between Iraq and France was a first round match in the FIFA World Cup. The match was played in San Paolo in heavy rain and strong winds. The two teams were evenly matched and both played open fast football with the ball crossing the centre line more than 30 times in each half. There were 21 shots on goal. Iraq scored in the 25th minute and France in the 45th and 81st minutes. The 2–1 result in favour of France has clearly established the strength of the teams and had a significant bearing on the World Cup outcome.'

- One sentence describes the research context – what is the general field of research and why is this particular research theme important?
- One sentence describes the research method(s).
- Two sentences describe the most important research results (including numerical results and the statistical strength of the results).
- Two sentences outline how these results and conclusions impact on the research field.

The process of writing the report implies that this part of the research has been completed. Therefore, logically, the sentences describing the research method and the results should be described using *past tense*. In the same manner, the discussion of previously published works should be expressed in past tense.

7.6 Paper preparation and review

There are many books, journal articles and tutorials which describe the challenges of writing a research paper [1–6]. In most scientific and engineering journals there is a page limit that must be strictly obeyed. Papers that exceed these limits can be automatically rejected or will attract a significant page charge for the over-limit pages. Most journals welcome complete papers which are shorter, even substantially shorter, than the maximum page limit. There is no advantage in adding additional information to a paper if this information is irrelevant or repetitive.

Scientific writing usually conforms to the common format with section headings listed as (see Figure 7.1):

1 Introduction;
2 Background/literature review;
3 Research methods;
4 Research results;
5 Discussion and conclusions;
6 Acknowledgements;
7 References.

It is advisable to use more specific headings and not the generic headings listed above. This reduces the need for some introductory sentences at the beginning of each section.

Example 7.7 Section headings

A paper on a water purification system using green sand might have the following section headings:

World challenges in water purification (= Introduction);
Common purification methods (= Background/literature review);
The green sand filter pilot plant (= Research methods);
Filter performance (= Research results);
Discussion and conclusions;
Acknowledgements;
References.

The page limit applies to the total content of the paper including title, abstract, all of the main text, the references and all figures and tables and their captions. This means that a writer must pay particular attention to the proportions of each section. There is no requirement to have sections of the same size, but although authors may write sections of significantly unequal length, a simple calculation of the approximate size for each section can be used as a general guide.

Example 7.8 Section length estimation

If the page limit is four pages with two columns on each page, and a standard format is used, then one can use the following section limits as an estimate of how much to write.

Title and abstract (0.5 columns), references (0.5 columns), six figures with captions (two columns, and two tables with captions (0.5 columns). This leaves space for 4.5 columns for the sections covering the introduction, background/literature review, theory, experimental/research methods, results, and discussion and conclusions. On average each section will have 0.75 columns of words and equations. Clearly the descriptions must be concise and directly to the point.

It is essential that all figures and equations follow the rules:

- Figures are referenced in the main text.
- Figures have captions which are self-explanatory (i.e. do not require the audience to read the main text to discover the meaning and importance of the graph).
- If there is more than one line on the same plot then these lines can be distinguished using a legend or other labels on the graph.
- All equations must be numbered.
- All symbols in the equations must be defined in the text close to the equation where the symbols are first used.
- All symbols must be unique, that is, a symbol must never be used to represent more than one quantity or variable.

For printed papers, colour pictures require special processing and so might incur an additional expense. For this reason, writers should attempt to use black and white images. This means that multiple lines on the same plot must be distinguished using different line styles (dots, dashes, continuous) and different symbols (e.g. *, o, +, etc). For papers published on the internet alone, there is no additional cost for coloured images as there is no requirement for colour printing.

Some journals require that authors submit their papers using a template. This is called 'camera ready' and means that the paper will be submitted in the correct style for publication in the journal on the first submission. In this case the authors must take great care to meet specified style requirements. Papers submitted in this way are usually not required to have page numbers as these will be added later by the journal editorial staff.

Other journals require the submission of the title, abstract, authors and keywords on a web page, and the main text, every figure and every table are submitted as separate files. Once all sections have been validated (i.e. the picture files are numbered correctly and all have been checked for clarity and the captions have been added on-line), the paper is assembled as a single *.pdf file. This file must be reviewed and checked by the authors before final submission to the editorial board is possible (all of these processes and checks are automated in on-line paper submissions).

The Editor-in-Chief of the journal will allocate an Associate Editor who will seek the (voluntary) services of two or more reviewers based on their expertise. Some journals require the authors of the paper to suggest the names and contact details of suitably qualified people who can be approached to review the paper. Once the reviewers have accepted their nomination, the file is sent to the reviewers for consideration. In a 'blind' review process, the names and affiliations of the authors are not included in the file sent to the reviewers. The aim of this process is to reduce the possible prejudicial effect of the nationality, the laboratory prestige, etc on the reviewers.

The reviewers are required to comment on both the technical content and the quality of the language used, and will submit one of the following recommendations:

1 Accept (no changes required);
2 Accept (minor editorial changes required);
3 Accept (major changes required); or
4 Reject (usually because there is no major scientific or engineering advance reported, or parts of the paper have been previously published, or parts of the paper have been directly copied, or the paper is technically incorrect).

The reviewer's decision will be justified by some general comments on the paper. The Associate Editor will review the reports from each of the reviewers and then make an overall decision in one of the four categories above. The Associate Editor's decision and the reviewers' responses will be returned to the authors for information. If a revision is required, then the paper must be corrected and resubmitted together with a detailed description of what changes have been made. All information will be returned to the reviewers and the process repeats until a final decision is made to accept or reject the paper. The review process will be undertaken once, twice or even three times, depending on the reviewers' comments and the Associate Editor's decision. Usually all reviewers of the paper must agree that the paper is acceptable for publication before the Associate Editor will recommend publication.

Following the acceptance of the file, the journal editors will arrange typesetting the paper (if not camera ready) including the figures and tables, correct any typographic errors and send a copy of the paper back to the authors for verification. If the paper has already been typeset (i.e. a template has been used and verified electronically), then only minor changes to the language and references will be required. After a copyright form has been signed by the authors transferring the copyright of the paper to the journal publisher, the paper will be officially published on paper and/or on the web and will become available widely to the engineering community.

7.6.1 **Paper reviewers**

The search for and selection of reviewers is a vital part of the peer review process, and the authenticity of the scientific method. Reviewing is a task which should not be undertaken lightly or by those inexperienced in the field. It is important that reviewers only accept the task if they are competent and knowledgeable in the specific field.

The specific instructions to reviewers relate to making comments and recommendations on the following topics:

- *Abstract*: Does the abstract sum up the major findings clearly and succinctly?
- *Novelty*: Does the paper present new knowledge which is important to the engineering discipline?
- *Literature review*: Does the paper adequately refer to the most important published literature – both historical and recently published articles?

- *Experimental methods*: Is the experimental method appropriate and is the method capable of yielding the correct results to the accuracy required?

- *Results and analysis*: Is the method of analysis appropriate and statistically valid? Have the results been independently verified through a comparison with the work of others or through other methods – theory, computational modelling, etc?

- *Discussion and conclusion*: Do the conclusions logically follow from the research results? Have the authors reviewed their methods and commented on the strengths and weaknesses of their experimental method? Have the authors commented on the next logical steps of this research? Have the authors commented on the impact of this work on the engineering discipline?

- *Quality of writing*: Is the language clear and concise? All typographical errors must be noted and referred back to the authors.

- *Quality of the figures*: Are the figures sufficiently clear? Are the axes on the plots labelled correctly together with units? Are the captions self-contained?

In undergraduate teaching, university academics are required to mark student laboratory reports and assignments. The feedback given to students through this process is vital to the education of the next generation of engineers. It is important that under-graduate students write their reports in the same manner as a scientific or engineering research paper. It is then very important that the academic marking those reports conducts the review of the submission in the same manner as (s)he would review a paper submitted to a research journal. Thus a junior academic can gain experience in the process of reviewing papers from undergradu-ates as the performance and the knowledge of the students grow proportionally.

After finalising a research paper, a team of novice researchers should apply the criteria given above. If the research team is uncertain about the quality, then it is more than likely that the reviewers will take an even stronger negative view and the paper will be rejected. This is self-check on the quality of a paper about to be submitted for review.

7.6.2 Paper review responses

When a paper has been submitted for publication in a journal, and the authors receive the feedback from the Associate Editor as 'minor changes' or 'major changes', they must make corrections to the original submission in a fixed time allotted by the journal editorial committee.

Commonly the paper will be resubmitted to the journal and sent out for review to the previous reviewers. In order to make

Example 7.9 Response to reviewers' comments

Reviewer comment 1: The authors have not adequately addressed the previously published work of Jones *et al.* (ASCE Section E vol 12 pp. 27–36). This paper reports a similar experimental method.

Authors: We had not seen this paper before. It does demonstrate a similar technique but the results were applied to a very different soil type. We have now included three sentences in the literature review describing their technique (see the highlighted section in the revised document).

Reviewer comment 2: Figure 5 has incorrect units on the y-axis. Please check.

Authors: The figure has been changed and the revised axis label is now correct.

Reviewer comment 3: Do the authors think that their technique can be applied to all soil types (as claimed in the conclusions)? I believe that the presence of water in the samples will prevent accurate results from being obtained.

Authors. This is correct. We have revised that sentence and included the phase 'for predominantly dry soils'.

etc

the second review relatively simple for the reviewers, the authors must submit two documents:

1 The revised paper with the important changes highlighted; and
2 A list of all of the reviewers' comments and the author's response to those comments.

7.7 Conference presentations

7.7.1 Preparation

The presentation of research results at a conference is an important method of engaging with the research community of peers. Each conference will have a Technical Committee responsible for the assessment of papers submitted for presentation at the conference. This process occurs in the time frame between six months and one year before the conference is held. The guidelines for papers (Figure 7.1) apply equally to the submission of a conference paper. A complete title and an abstract are required in a form similar to that described in Sections 7.4 and 7.5.

Commonly the presenter will be required to use a computer set of images (called 'slides') projected using a digital projector. MS Powerpoint (*.ppt and *.pptx files) and Adobe *.pdf files are the most common files used in conference presentations. Note that Powerpoint provides guidance about layout and suggestions for uniform appearance in all of the slides through a master slide template that can be edited.

Conferences are organized and run with the speakers given strict time limits for their presentation. Each conference session is chaired by a person who is responsible for the timely presentation of the papers allocated to the session. Each speaker is allocated a time slot (start time, presentation end time and question time). This information is well known before the conference and must be followed precisely by both the speakers and the session chairs.

The structure of the presentation should follow Figure 7.1.
There are some general presentation rules (guidelines) which
can be used to estimate the time required for the presentation
[2, 4, 5, 6].

1 The presenter should never plan to present slides more rapidly
   than one slide each minute. A presentation faster than this
   makes it too difficult for the audience to follow.
2 Every slide should be clearly numbered. This allows an audience
   member to indicate which slide is to be discussed during
   question time.

Example 7.10 **Presentation timing**

A talk is scheduled for ten minutes presentation time with
five minutes for change over to the next speaker. The total
slide count for this time allocation should be a maximum of
13 slides as follows:

The first slide is the title slide and will require little explanation apart from acknowledging the co-authors of the
paper.

The final two slides with be a *thank you* and a list of
references and acknowledgements. These two slides will
not require verbal comments during the presentation.

Thus the presentation must be delivered in full (introduction,
theory, research methods, results) in ten slides. Assuming a
maximum of two figures in one slide, the schedule should be
something like this:

Introduction (1 slide);
Theory (1–2 slides);
Experimental method (1–2 slides);
Results (3–5 slides);
Discussion and conclusions (1 slide).

3   The number of words on each slide should be less than 50 and the text should be broken up into fewer than eight bullet points.

4   Animated images must be checked on the computer and the display used at the conference venue. There are many compatibility problems with different computers and projection facilities.

5   References must be included in the presentation at the end of the talk as a full reference and/or on the relevant slide as the name and date only.

6   Slide colours are very important for clarity. The presenter should choose a background with a light neutral colour and clearly contrasting letters and equations. Preferably the auditorium/lecture theatre should remain well lit so that audience members can take notes. A dark environment prevents note taking and is also conducive to sleep and so should be avoided.

7   The font type and font size must be chosen for maximum clarity. This usually means a font size of greater than 20 point should be used.

7.7.2 Delivery

Before the day of the presentation, the speaker should run through the presentation out loud, alone or in a small group, to ensure that all concepts can be clearly explained and the talk can be delivered in the time allocated. This also gives the speaker the opportunity to form sentences. This skill is often overlooked when reading the presentation slides.

There are some simple methods to minimize stage-fright for speakers that are nervous. These include:

1   Rehearse the talk a number of times before the event with and without an audience.

2   Memorize the first two sentences that will be spoken. This allows the speaker to start without being too hesitant.

3   Use hand note cards. This is an insurance policy. Rather than using these hand notes during the talk, they can be used to

rescue the speaker from nervous confusion. Commonly, however, the slides on the screen can do this more easily. MS Powerpoint allows the speaker to see notes next to the slides on a local monitor without being revealed to the audience. This is a simple electronic replacement for the hand note cards.

4 Make sure each slide has a requirement for the speaker to explain something more than the written text. (Talks that are delivered by reading the screen or reading a pre-prepared paper are usually boring for the listeners.)

5 Do not commence speaking to the presentation until the title slide appears on the screen and the Chairperson of the session has introduced you.

Conferences are usually organized with a timetable/schedule that allows the speakers in a session to load their presentations from a memory stick (USB thumb drive) and to ensure that they open correctly. (Note that it is best to name each file with the presenter's name rather than the conference name as there will be many presentations loaded onto the same computer. Commonly speakers are not permitted to use their own computers as technical faults can arise and these can cause delays.) This should happen before the session starts: usually during a coffee break or lunch break. It provides an opportunity for the speakers to meet the session chairpersons. It is important that the session chairs know that all speakers have arrived and have successfully loaded their presentations into the computer. This is also confirmation that the speakers are in the correct room and at the correct time.

During the delivery of the talk, the speaker should face the audience. The sound level of the presentation should be judged by imagining that the presenter is talking to the people seated in the back row of the audience. This should ensure that the voice projection is adequate. The speaker should attempt to speak slowly in short, simple sentences as the language used might not be the speaker's native language, and might not be the native language for many members of the audience. The explanation of

Chair's responsibilities (a rough guide)

Before the day of the session
Read the papers in your session
Before the session starts
Meet the speakers and assist with presentation loading, the projection equipment, pointer, etc.
Before the talk
Introduce yourself (name and affiliation only) and start the session on time
Introduce each speaker at the appointed time (Learn the correct name pronunciation)
During the talk
Indicate to the speaker when there are 2–3 minutes to go (usually by fingers in the air)
Politely stop the speaker when their time has expired.
Assist the speaker if problems arise with AV, external noise, etc
After the talk
Thank the speaker
If there is time, seek questions from the audience
If there is time, ask questions of the speaker (to ensure there is not silence during the question time)
At the end of the session
Thank all the speakers and the audience.

FIGURE 7.2 The roles of the session chair at a technical conference.

the presentation material should be done verbally and preferably spontaneously (i.e. without looking at notes).

When presenting graphs, the axes must be mentioned before commenting on the data in the graph. When presenting equations both sides of the equation should be explained verbally and the important symbols defined.

It is important that the talk has a clearly defined end slide. This is achieved using a slide which has the two words *thank you* only (perhaps with some images for colour), and the speaker should thank the audience for their attention. This signals to the session chair that the talk is finished and s/he should invite questions if time permits. The speaker should not invite questions. This is the role of the chair of the session only.

The roles and requirements of the session chair (Figure 7.2), and the speaker (Figure 7.3) are well defined. It is essential that the speaker does not exceed the time limit for the talk. To talk beyond the allocated time is extremely rude and offensive to the chair (who has the responsibility to stop the speaker on time), to the following speakers (is their presentation less important that the current presentation?) and the audience (who wish to move to another room to listen to another paper timed to begin at the end of the current paper, or take a coffee break/lunch, etc). Many

---

Speaker's responsibilities (a rough guide)
Before the day of the session
Prepare the presentation
Ensure the length of your paper fits the allocated time
Before the session starts
Introduce yourself to the session chair before the session starts
Ensure the presentation is loaded, is error free, and is readily available (computer desktop)
Learn to operate the projection facilities
The talk
Briefly introduce yourself and your co-authors (names and affiliations)
Thank the session chair for your introduction
Speak clearly and loudly (speak to the back row in the room)
Finish on time by clearly stating *Thank you* to indicate you have finished your presentation
After the talk
Wait for the session chair to call for questions
Answer the questions directly and concisely
After the session
Return to your seat and wait for the session to finish
Be prepared to stay after the session for further discussion of your paper

---

FIGURE 7.3 The roles of the speaker at a technical conference.

conferences have more than one session running simultaneously in different rooms. Large international conferences might have 18 or more parallel sessions running at the same time so moving between the presentation rooms can take some time.

Example 7.11 **The firm session chair**

A session chair had a challenge with one speaker. Three minutes before the scheduled end of the presentation the chair (seated in the front row facing the speaker) indicated there were three minutes to finish. At the scheduled finish time the chair waved politely for the speaker to stop. The speaker did not. After two minutes into the questions/changeover time, the chair stood up and moved close to the speaker. The speaker continued the presentation. After another minute the chair put his hand on the speaker's shoulder and spoke into the microphone and said *thank you*. The next speaker was introduced. The verbose speaker was forced to stop without questions and without finishing.

Novice and experienced presenters must not speak beyond their time allocation.

7.8 Poster presentations

Poster sessions at science and engineering conferences can be of significant benefit in providing the authors with an opportunity to engage in the event with enthusiasm. It is an opportunity to present research work one-on-one to interested parties. Unlike conference presentations, when question time is very restricted, a poster session allows longer discussions of the merit and quality of the research work reported. This is one reason to choose to present the work at a poster session. Other advantages include the opportunity to discuss the research without suffering the stress of making a public presentation and the challenge of answering questions at a very public forum. This is particularly true if the authors are required to present in a language other than their native tongue. Often authors in this situation find it quite difficult to present and might fail to understand and respond to the questions at the end of the presentation. At poster sessions, an author can ask questions of the audience; this is not common practice during the question time following a presentation.

In a poster session at a conference, the authors are required to prepare a large poster; the A0 and A1 formats are common. During the poster session, the author is required to stand close by the poster to explain what research is being reported and to answer questions about the work, the proposed further work, etc. As the author might not be able to speak with all passers-by, and at times the poster might be unattended, the poster must be self-explanatory. The title must be sufficiently large so that it is

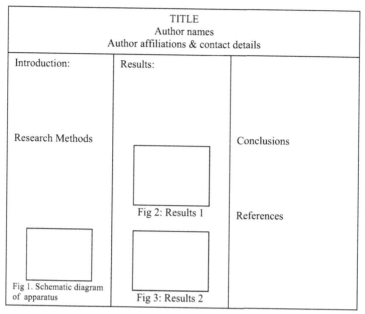

FIGURE 7.4 Typical poster layout in landscape. The proportional area dedicated to each section and the number of columns are usually not rigorously defined, but the font size should be sufficient so that a small group of people can read it simultaneously from a distance of 1–2 metres.

readable from a distance. The general layout of a poster is given in Figure 7.4. The font size and the figures need to be sufficiently large so that a small group of people can read and understand the information at the same time.

During a poster session at a conference, authors are required to mount their posters before the session starts, and to be available to discuss their poster during scheduled viewing sessions – often over a period of one day or one half day. The necessary ingredients for a successful poster session include an enthusiasm for the subject and a good understanding of the underlying theory of the project.

To maximize the opportunity presented at a poster session, the authors should be ready to:

1   Give a 3 minute outline of the project and the main results;
2   Answer questions;

3 Give additional important information about the research more generally (e.g. published paper on the topic, business cards and other contact details for further discussion after the conference has finished).

7.9 **Patents**

When patent protection is required for research outcomes, the submission of a patent application must take place before the submission of a paper or conference presentation. Patent writing is often undertaken by hired professionals rather than the researchers themselves because the process has a legal framework and does not generally follow the structure outlined in Figure 7.1.

The key element of a patent is the list of 'Claims'. The claims are a series of succinct descriptions of what is new, i.e. the technical advances made in the description of the work and its outcomes. These claims will form the basis of the patent assessment and so must include sufficient information and qualifying sentences to ensure the claims are unique.

In writing a patent there is significant emphasis on referring to all relevant patents in addition to the openly published literature on the topic. A patent will only be granted on the basis that either the idea is novel or that no-one with expertise in the area could deduce the idea from a knowledge of previous patents and other published work. This can be quite difficult for authors who have little experience in patent writing and patent searches. A patent can be filed without experimental verification. It is the concept only, as described by the claims, that is most important.

7.10 **Chapter summary**

Report writing, publication and presentations are the final requirements in the research process. A failure to engage in these activities might mean that the research outcomes were not successful, not significant, and perhaps fundamentally flawed. These views lead to the conclusion that publication and/or presentation of the research outcomes are essential. For more commercially oriented outcomes, publication in the internationally reviewed literature validates them and provides independent support for the commercial development of devices, strategies, databases, as well as confirmation that the work is appropriate for inclusion in an engineering standards document.

Patent writing is a process requiring technical skills in the profession. For this reason most patents must be written and submitted using patent attorneys before journal papers or conference presentations are submitted. If the patentable idea is presented in an open forum, it can no longer be patented.

The writing and presentation techniques presented in this chapter are simply guidelines for novice researchers. While the basic elements remain essential, the method of approach can vary in line with the publication methods common to a particular research field, and the flexibility afforded to more experienced researchers. The best guide to writing papers and abstracts is the published work in the field, and the specific journal in which the paper is to be submitted. There is a preference held by some journals to ensure that relevant work previously published in that journal is referenced in the paper, however, this should not be mandated.

Example 7.12 **Journal referencing**

Journals are sometimes rated by their impact factor. The impact factor is based on the number of citations of papers from that journal. Therefore the editorial staff of a particular publication are keen to gain maximum benefit from all new papers that are to be published in the journal. One method is to include many references from that journal in the new papers published.

This pressure must never override the important concepts of researcher independence and critical review.

The format of all written (journal papers) and oral presentations (conference paper presentations and poster presentations) is very similar. The pattern follows Figure 7.1 where the start and end sections should explain the context in the wider discipline. The central sections must be focused on the theory, experimental and analysis details.

Almost all papers and presentations require an abstract. The format of the abstract is similar to that shown in Figure 7.1 but with a very limited word count. Most important is that the abstract should clearly state the key results, preferably in numerical form with a statistical assessment of the uncertainty.

There may be restrictions on researchers wishing to submit a journal paper following successful presentation at a conference. In this case, the copyright assignment for the conference paper must be checked to ensure further publication is allowed. It is wise to change the wording and enhance the content of the journal paper in order to reduce accusations of plagiarism and subsequent paper rejection. These issues are covered in Chapters 1 and 2.

Example 7.13 **Report publication**

You write a short paper for a conference in which you are to report ten events/experimental results. The conference paper is accepted on that basis. In the meantime your research group has continued the research work. By the time of the conference the research team has 250 test cases to report with much better statistics. You write a much stronger paper for a journal and submit it for publication. You must ask the question: 'Have I copied too much from the conference paper?'

The method of approach in the journal paper is to reference the conference paper explaining the number of samples and general conclusions. Where appropriate the journal paper should have an expanded experimental method and of course, much improved statistics. The conclusions can be written in such a manner as to support the brief conclusions given in the conference paper. This is a valid method of approach as the conference paper will be published a long time before the journal paper. The journal reviewers can review both the conference paper and the new journal paper and make a decision on the new content. It is important that all text and figures are not identical to what is published in the conference proceedings.

The review process takes some time between the submission and the return of the reviewers' reports. The reports must be addressed in full and in sufficient detail to ensure the reviewers gain an understanding of the changes made to the document in the light of their recommendations, and the reasons why some recommendations have not been addressed if a good argument can be presented.

The research process is not complete until the papers have been accepted by a journal or conference and the peer review has ascertained the validity of the work and the significant contribution made to the field of research.

Exercises

7.1 Ask a colleague to take a published journal paper in your discipline and remove/obscure the title, authors and abstract. You are then asked to read the paper. You are now required to write a new title and abstract of this paper using 100 words, based on what you believe are the important issues in the paper. Once complete, compare your paper title and abstract with that in the original papers. Is your submission better or worse? Explain the reasons for your views.

7.2 Take a published paper in your field and create a poster using direct cut-and-paste of the words, title and diagrams. Is your poster self-explanatory?

7.3 Using the same paper as in Exercise 7.2, prepare and deliver a five minute talk as you might at a conference. You should be prepared to deliver the talk within the time limits and answer questions from the audience.

7.4 Assume several members of a research laboratory have papers accepted at a conference. Arrange a mini-conference using the timing, etc defined by the conference. Ask an experienced researcher to chair this session and complete the process in full (i.e. to introduce the speakers, time the paper, give a two minute warning before the end of the presentation time, thank the speaker, lead the applause, host the questions and then move on to the next speaker).

7.5 Design a two minute paper presentation. Prepare one slide (only one!) showing a plot of the major result(s) from a research project. Deliver the two minute presentation on the topic showing only one slide of results (no title slide and no conclusions). This is good training to develop clear presentation techniques.

7.6 Read a journal paper from your discipline. Following the format of patents, write out one or more important outcomes from the paper in terms of one or more Patent Claims 1, 2. These claims must not only be new, they must be not-obvious from previous work.

References

Keywords: report writing, publication guidelines, instructions for authors, scientific presentations, patent

[1] Katz, M.J., *From Research to Manuscript. A Guide to Scientific Writing*, 2nd edition, New York, NY: Springer.

[2] Booth, W.C., Colomb, G.C. and Williams, J.M., *The Craft of Research*, Chicago: U Chicago Press, 2008.

[3] Lindsay, D., *Scientific Writing = Thinking in Words*, Collingwood, Victoria, Australia: CSIRO Publishing, 2011.

[4] Snieder, R. and Larner, K., *The Art of Being a Scientist. A Guide for Graduate Students and their Mentors*, Cambridge, UK: Cambridge University Press, 2009.

[5] Alley, M., *The Craft of Scientific Presentations: Critical Steps to Succeed and Critical Errors to Avoid.* New York: Springer-Verlag, 2003.

[6] Hofmann, A.H., *Scientific Writing and Communications: Papers, Proposals and Presentations*, New York, NY: Oxford University Press, 2009.

8

The path forward

A research project is about creativity. A research team can only exercise this creativity if they:

- Know and understand their engineering discipline well;
- Read the literature and continue to keep up to date with recent publications;
- Maintain cordial relationships with colleagues and funding organizations;
- Are honest and ethical in all aspects of the research process;
- Plan and remain well organized;
- Stick to the schedule if at all possible.

While the motivation to undertake research must be primarily to improve the human condition through the development of new engineering products and services, there are significant personal rewards for researchers through a published legacy of achievement which leads to reputation building and employment opportunities.

The engineering disciples continue to grow and change. So too do the methods of interacting with the general public and the application of new technologies to the research process and publication of results. As these new processes and technologies arise, professional engineers must maintain their ability to engage with new technology as well as to continue to contribute to these changes through published research.

8.1 Publication trends

In this book, heavy emphasis has been placed on journal publications, both to present the research outcomes and to maintain current knowledge of developments in the relevant engineering discipline.

In this digital age there have been some rapid changes in the cost of access to publications – even a researcher's access to his/her own published works. The publication of journals incurs some costs for the editorial staff, the printer (if published on paper) and internet costs. The costs models outlined in Section 2.7 continue to evolve as professional societies attempt to cover costs and commercial publishers attempt to make a profit for their shareholders. For professional societies, normally there is no 'profit' component to the cost structure and papers might be published without cost to the authors or at a relatively small cost. Of course the society must make sure that publications are not a drain on the society's accounts. There must be a break-even point for a not-for-profit publisher where the cost of publication is balanced by revenues obtained through the sale of society membership, conference registrations and journal articles. The commercial publishers seek to make a profit and the costs might be significantly higher, even if the editorial work is conducted by volunteers who are members of a professional society.

At present there are a number of methods of gaining access to publications:

(a) A fee for access will apply to individual papers in most traditional journals for someone who is not a member of the professional society which hosts the publication. Members of professional societies usually have access to their society's journals as part of the membership fee (sometimes at an extra cost in the membership package). Research institutions might subscribe to particular journals (or groups of journals), but these subscriptions can be very expensive indeed. Researchers based in a research institution should check which journals are available through the institutional subscription before paying the fee for a particular article.

(b) Some journal publishers allow the author to freely distribute papers they have authored for scholarly use. Some web search engines will note that a particular paper is available without cost.

(c) Some on-line journals allow access to all published papers without cost. These journals will normally charge the paper authors a fee for publication.

(d) Collections of journal papers (for example one year of papers in the journal) can sometimes be purchased. Many societies will sell the complete archive of all papers in a journal. Conference proceedings (the complete collection of papers) usually have an ISBN number. It is sometimes possible to purchase one or more volumes of conference proceedings.

8.2 Getting started in research

This book has described the fundamental requirements for undertaking engineering research. Clearly, even a relatively short project requires a research plan, a research team, an approval from various organizations, followed by reporting. Some strategic approaches are suggested in Section 3.2, where the novice research team is encouraged to develop some research ideas and a research project from previously published research.

Another alternative is to consider writing a commentary on a previous paper or group of papers. There are two types of journal papers which do not require new work to be undertaken before publication. These are classified as 'Comments' and 'Reviews'. In both cases, the researcher must have a comprehensive understanding of the engineering discipline and a strong, current familiarity with the relevant literature. This allows the research team (often an individual researcher) to make a useful contribution to the research field. As before, the research team must decide on a research question before writing the paper.

8.2.1 Literature reviews

In the case of a review of the literature, the research question might take a form such as:

> *Can accelerometers estimate velocity in human movement?*
>
> *Do battery power and size follow a form of Moore's law?*

These types of questions can be very important to the research community. The method of addressing these questions is to review the published data and formulate a statistically strong argument based on the work of other groups around the world.

Example 8.1 **Review paper**

A typical review paper would include a general introduction to the topic followed by a discussion of various techniques that have been applied to addressing the problem. These techniques might be divided into different categories such as:

Theoretical analysis;
Experimental measurements;
Numerical modelling;
Case studies.

The paper is summarized with a conclusion about the status of the research and a discussion of the unsolved problems.

Review articles are often a defined category of paper during the submission process. There may be different requirements in relation to length, number of references, etc. If the article is to include figures, then the necessary copyright requirements must be met before the article is allowed to be published. Copyright permissions often require a payment and an indication to the copyright holder of the type of publication where the image or words will be used. The process of obtaining copyright permission can be lengthy.

A review article will be subjected to the same peer review process as other research papers.

The presentation of a review paper at a conference is more likely to be in the form of an invited paper by an eminent person in the engineering discipline. Some papers would be delivered in a plenary session at the conference. Unsolicited review papers are unlikely to be accepted, as the major motivation for arranging and attending a conference is to present new developments in the field.

8.2.2 Comments

There is merit in making a comment on a recently published paper (most journals require the comments to be published within a year of the date of the original publication). In this case the research question might take the form:

Has the theory developed by this paper been adequately tested?

Can the results of the authors be reinterpreted to support a different theory?

Is the research method sufficiently robust to support the conclusions given?

Scientific controversy is very important in the development of science and engineering. Many very famous discoveries were initially rejected by the reviewers of the first journal paper submissions. For this reason, short papers that comment on the scientific method and conclusions are usually encouraged. Of course personal attacks on the authors of a paper will not be published.

Comments published in journals are limited to a small number of words. No new concepts should be introduced so that this technique cannot be used to develop new theories or report additional experimental outcomes. A comment must review the paper and provide a well-grounded scientific basis for a different interpretation of the data. As with all scientific papers, there must be appropriate references to the literature. It is not adequate to claim that some unpublished results disagree with or contradict the findings unless previously published works can be cited. If some contradictory findings have already been published in the open scientific literature, then there is a strong argument to submit a comment for publication.

When a comment is submitted to a journal, it is forwarded to the authors of the original paper for a response. If the editor approves of the discussion, then both the comment and the response will be published together. In this manner questions of fundamental importance can be addressed without further experimentation.

Example 8.2 **Comments and discussion**

The fundamental electromagnetic performance of a staked antenna used in ground probing studies for mining, geotechnical and geophysical applications was the subject of a series of comments and replies [1–4]. The discussion involved several exchanges between the two parties.

8.3 Quality assurance (QA)

The delivery of products and services error free and on time is guided by an adherence to quality standards [5–7]. Engineering companies commonly comply with the international standard ISO 9000 and those companies wishing to gain certification to this standard must implement processes outlined in ISO 9001. An organization is required to maintain a *Quality Policy Manual* and an *Operational Procedures Manual* in order to comply with the standard.

While research is a discrete, time limited process, there are many advantages to be gained in using quality assurance techniques in achieving the best possible research outcomes. Some of the techniques suggested in this book form part of the QA requirements in organizations. For example the use of statistical techniques and standard research and measurement procedures improves the reliability of the research outcomes.

Many useful concepts from QA methods will improve research outcomes. These are briefly described in the following subsections.

8.3.1 Record keeping

There is an obligation on researchers to keep a log book in which all experimental planning, methods, and results and conclusions are noted. This is particularly important if the research outcomes are to form part of a patent, but also to ensure that cases of plagiarism and inappropriate data manipulation can be traced. Through appropriate sign-off by the researcher and independent observers on a regular (daily or weekly) basis, the log book constitutes legal proof of the date and content of the research developments.

From a QA perspective, a record of calibration procedures and other instrument maintenance requirements serves to verify the accuracy of the research outcomes.

The sharing of data within the research group ensures reliable record keeping – even in the event of data loss by one person. Commonly data loss occurs through computer storage failure but also paper records can be destroyed through fire or flood.

The research team must maintain a logical method of version control for all documents and computer programs. This ensures that the data analysis methods can be repeated even several years after the initial research work was completed.

Example 8.3 A revised research theory

A research team conducts a series of geotechnical experiments in the vicinity of an active volcano. After the measurement data have been gathered, the team analyses the data following a particular theory popular at the time.

Five years later a new theory emerges and can be tested using the same data set. By this time the measurement region has been destroyed by the volcano. The research team can reunite and process the data again using the new theoretical approach. Excellent record keeping will ensure that this re-processing of the data is possible.

8.3.2 Sample storage

The fabrication or collection of samples can be an expensive and time consuming procedure. Commonly more than one sample is used in the research to improve the statistical support for the research conclusions. The method of storage of the second and other samples should be undertaken in a rigorous manner to ensure that subsequent analysis is possible.

Example 8.4 **Duplicate samples**

Soil samples, concrete samples, semiconductor samples, ice cores from the Antarctic and thin films will all degrade if stored inappropriately. Some will require storage in an oxygen-free environment, others will require a specified temperature and humidity level to prevent environmentally induced change. The research team must develop and adopt a protocol which ensures the samples are stored in a manner in which this degradation is minimized.

8.3.3 Staff training

When developing a research proposal, the research team must consider involving external experts who can reliably undertake small but vital aspects of the research. In some cases this might be

Example 8.5 **Research training**

A new and complex test apparatus is purchased as part of a research grant. The supplier offers a training course for two members of the research team at a small additional cost. Engagement of staff in this training will increase the speed at which reliable use of the machine is possible and the best possible measurement outcomes will be achieved. The lifetime of the instrument can be extended by appropriate training in the use, calibration and servicing.

Short training courses on materials handling, instrumentation methods and data processing standards will ensure improved research outcomes.

too expensive or too time consuming. For this reason, the training of research staff is a very appropriate method of ensuring the best possible research outcomes.

8.3.4 Design control

The development of new instrumentation will usually be quicker and more reliable if the tools used are familiar to the research team. While there might be a temptation to engage in using the very latest technology, the speed of development might be slow and the final outcome might be inadequate if the new technology is not well understood by the team.

Example 8.6 Equipment decisions

The research team wishes to develop a set of wireless sensors for river monitoring. The electronics technician suggests two options: (a) the use of a new integrated circuit which includes analog inputs, large memory, many serial ports and simple programming; and (b) a set of integrated circuits with separate functionality. This chip set was used successfully to develop the last three networks.

What is the better decision for the research team?

8.4 Occupational health and safety

There is a legal obligation in addition to an ethical obligation that all research work is conducted in a manner in which there is no harm to the researchers, other people in the institution and the general public at large. The risks should be documented in a risk assessment spreadsheet in which all possible risks are outlined and methods minimizing those risks are explained and implemented.

This means that people with access to the research site/ laboratory must be protected or trained to ensure that the risks are minimized. There must also be a display of the contact details of medical emergency personnel and a fully stocked first aid kit.

Example 8.7 Equipment safety

A new high voltage, high pressure machine has been purchased and installed in a laboratory. It is essential that unauthorised and untrained people do not have access to the machine. This can be achieved through physical barriers, key-locked switches, and the requirement for multiple users to start the machine.

The chemicals used in the laboratory must be stored safely and the hazard sheet for every chemical must be immediately available to all persons in the laboratory.

Experimentation outside of the laboratory also requires that the safety aspects are addressed. Hazards to the researchers include falling from heights, impacts from vehicles and other moving objects and exposure to the sun, rain, wind, heat, etc.

Example 8.8 Field research safety

The research plan involves making measurements on an active road bridge. The researchers in the field must be equipped with fluorescent vests, steel-capped boots with non-slip soles, head protection from the sun and portable barriers and signs for oncoming traffic to slow down. Notification of the local road authority and the traffic police might also be required.

8.5 A glimpse into the future of engineering research

Research is often an internationally based activity with research groups around the world sharing expertise, resources, infrastructure and equipment. The development of on-line access to test and measurement platforms is a likely trend. Already autonomous machines are used in dangerous environments (e.g. space vehicles, bomb disposal, nuclear power plants, etc). Currently engineering companies employ remote control of such major pieces of equipment as underground autonomous mining vehicles [8], distributed power generation systems [9], robots [10, 11], manufacturing facilities [12]. There are many such examples of remote and distributed control technologies. It is eminently feasible that research equipment can be operated remotely or co-operatively. Remote sensing using wireless sensor networks has many applications in engineering research [13–15]. Unmanned aerial vehicle (UAV) technology is now capable of video streaming from above the earth's surface and even inside large buildings. UAVs can be equipped with sensor systems that can be used indoors and outside with precision, collision avoidance and co-operatively for long periods of time [16].

As new publication methods and access techniques develop in the future, all will require a cost structure that is neutral (break even) or profit based. As the perceived value and reputation of individual journals are scored by indices such as impact factor, each publisher will maintain a balance between open access (resulting in more journal citations) and cost-based access (to maintain the financial viability of the publisher). Publishers also face the challenges of minimizing the time between submission and publication without diminishing the validity of the peer-review process. It is anticipated that the use of social media running on highly-portable electronic/computing platforms will increase. There is scope for increased automation in the preparation, typesetting and publication of articles. The incorporation of multimedia presentations (e.g. moving images) and extensive databases into electronic publications is a reality with some publications. The development of new computer codes is a valuable contribution to many engineering disciplines. Some journals are happy to provide such codes for all users to test and verify.

Exercises

8.1 Writing a comment: Using your discipline's keywords find a paper that has been published within the last year. Review the paper and evaluate the strength of the arguments and the statistical analysis, compare the method with common practice and so write a 'comment' on the paper. Make sure that your arguments are backed up by other publications. This should be prepared in a style for publication using less than 800 words. Share your comment with others for review and discussion.

8.2 Reviewing a review: Using your discipline's keywords find a review paper (simply add the word review to your list of keywords). Read the review paper and note the number of references. Estimate the average number of sentences used to comment on each paper (or group of papers) and compare the subject material of each of these with the proforma outlined in Section 2.5. Write out a suitable research question for this review paper.

8.3 Planning a review paper: Using the keywords from your discipline and one additional, highly specific keyword, conduct a web-based literature search. Assemble a list of more than 20 potential papers suitable for your literature review and divide them into 3–5 different categories. Use a textbook as a reference to define the terms to be used in your review and write a 2000 word review using these references.

8.4 Safety: Complete a safety audit on the laboratory in which you work. Check for the following: restricted access, protective clothing, chemical information, proper waste disposal, electrical safety, fire and smoke alarms, first aid kit, etc. Make some recommendations about how safety in this space can be improved.

8.5 Training: Review a published paper in your engineering discipline. List the research methods that you feel unqualified to perform. Conduct a search of available training courses which cover these methods and establish the cost and time requirements of this training.

8.6 New technologies: Write a review of possible applications of new technologies in your engineering discipline. In particular note the precision of such measurements with reference to published research papers and commercial specifications of off-the-shelf instrumentation. Options to consider include wireless sensor networks, remotely controlled robotics and UAV monitoring platforms.

8.7 Quality assurance: Review a paper describing the ISO 9000 requirements for quality assurance. Read a published research article and comment on the relevance of ISO 9000 attributes to the research reported in the paper. Do you think the authors might have improved outcomes if there was better compliance with ISO 9000? Explain your answer.

References

Keywords: review, comments, ISO 9000, quality assurance, health and safety, wireless sensor networks (WSN), unmanned aerial vehicles (UAV), robotics

[1] Wu, X.W. and Thiel, D.V., 'Electric field probes for electromagnetic sounding', *IEEE Trans GE-27* (*1*), 24–27, 1989.

[2] Wait, J.R., 'Comments', *IEEE Trans GE-27* (1), 23–26, 1989.

[3] Wu, X.W. and Thiel, D.V., 'Reply to comments 1 and 2', *IEEE Trans GE-27* (6), 790–792, 1989.

[4] Thiel, D.V., 'On measuring electromagnetic surface impedance: Discussions with Professor James R. Wait', *IEEE Trans. Antennas and Propagation*, 48(10) 1517–1520, 2000.

[5] Sayle, A.J., *Meeting ISO 9000 in a TQM World*. Hampshire, England: AJSL, 1991.

[6] Threlfall, J., *Beyond ISO 9000: Further Developments in Quality Management*, Strathfield, N.S.W.: Standards Australia, 1996.

[7] Hoyle, D., *ISO 9000 Quality Systems Handbook*, Oxford; Boston: Butterworth-Heinemann, 2001.

[8] Roberts, J.M., Duff, E.S. and Corke, P., 'Reactive navigation and opportunistic localization for autonomous underground mining vehicles', *Information Sciences*, 145(1–2), 127–146, 2002.

[9] Peças Lopes, J.A., Hatziargyrious, N., Mutale, J., Djapic, P. and Jenkins, N., 'Integrating distributed generation into electric power systems: A review of drivers, challenges and opportunities', *Electric Power Systems Research*, 77, 1189–1203, 2007.

[10] Fitzpatrick, T., 'Live remote control of a robot via the internet', *IEEE Robotics and Automation Magazine*, 6(3), 7–8. 1999.

[11] Baudel, T., 'Charade: remote control of objects using free-hand gestures', *Communications of the ACM*. 36(7), 28–35, 1993.

[12] Wang, L., Orban, P., Cunningham, A. and Lang, S., 'Remote real-time CNC machining for web-based manufacturing', *Robotics and Computer-Integrated Manufacturing*, 20(6), 563–571, 2004.

[13] Ilyas, M. and Mahgoub, I. (eds.), *Handbook of Sensor Networks*, Boca Raton: CRC Press, 2004.

[14] Buchli, B., Sutton, F. and Beutel, J., 'GPS-equipped wireless sensor network node for high-accuracy positioning applications', *Wireless Sensor Networks, Lecture notes in computer science 7158*, 179–195, 2012.

[15] Thiel, D.V. and Lisner, P., 'Sensor networks and microsystems: get smarter', Proc. *SPIE5649, Smart Structures, Devices and Systems II*, 345, 2005.

[16] Merino, L., Caballero, F., Martinez-de Dios, Ferruz, J. and Ollero, A., 'A cooperative perception system for multiple UAVs: Application to automatic detection of forest fires', *Journal of Field Robotics*, 23(3–4) 165–184, 2006.

Appendix A: Matlab plot functions

Matlab® is a mathematical processing software suite which can be used for both discrete and continuous (symbolic) data sets. While initially based on a set of matrix mathematical functions, the program now includes a wide variety of functions covering many commonly used scientific and engineering processing tasks and display applications.

The simplest command line for a 2D plot is:

plot(X,Y,'*');

where X is an array of x-axis values and Y is an array of values to be plotted on the y-axis. This command line plots each discrete point using the '*' symbol and so is ideal for plotting experimentally derived data. The axes can be labelled using the functions:

xlabel('X parameter (units)');
ylabel('Y variable (units)').

The arrays X and Y can be imported from an Excel spreadsheet using the command line:

[X Y] = xlsread('filename');

or entered directly into the Matlab program as an array using the definition commands:

X = [1 2 3 4 5 6 7 8 9 10];
Y = [1.2 2.1 3.2 4.0 4.5 5.2 6.1 7.8 8.3 9.5];

where the numbers in the square brackets separated by a space constitute the data values.

A straight line between two points can be plotted using the following command sequence:

X=[Xmin Xmax];Y=[Y(Xmin) Y(Xmax)];
plot(X,Y);

where Xmin is the minimum value in the array X, Xmax is the maximum value in the array X, Y(Xmin) is the Y value at the minimum X value (Xmin) and Y(Xmax) is the Y value at the maximum X value (Xmax).

Once the plot is on the screen it is possible to edit some features of the plot including font size, data point style, line style, etc. The plot can be saved in image formats as stand alone graphics files or copied into other documents as part of a report.

Matlab has functions for plotting histograms and three-dimensional data using mesh, contour and image plots. Internet search engines will direct users to help files for these and many other Matlab functions.

Appendix B: Excel plot functions

The Microsoft Excel® program has many standard plot functions. As the program is icon based, the plotting of a 2D discrete data set requires the selection of the appropriate plot function (scatter). This function does not include interconnecting lines between the points (as required when plotting data points). It is necessary to highlight the x array and y array when prompted. Axis labels can be added during the formation of the plot.

Once the plot has been completed it can be edited by right clicking on an appropriate part of the graph (e.g. axis labels, data points, axis scale, etc). The plot can be selected, copied and placed into another document as part of a report.

Users are directed to the on-line help functions to assist in more complicated plotting operations.

Index

Printed in the United States
By Bookmasters